U0092228

商　用　英　文

程　振　粵　著

學歷：英國倫敦大學經濟學碩士
　　　世界藝文學院文學博士
經歷：曾任臺大、政治、東吳、輔仁
　　　等大學教授
　　　高等考試典試委員
　　　美國新聞處顧問
現職：亞盟秘書處顧問

三　民　書　局　印　行

網際網路位址 http://www.sanmin.com.tw

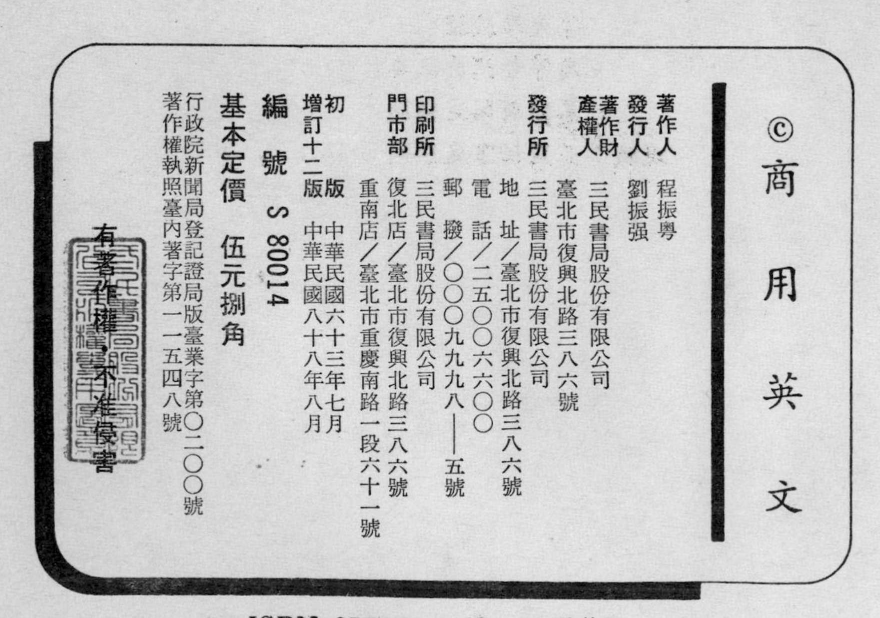

© 商用英文

著作人 程振粵
發行人 劉振强
著作財產權人 三民書局股份有限公司
臺北市復興北路三八六號
發行所 三民書局股份有限公司
地址／臺北市復興北路三八六號
電話／二五〇〇六六〇〇
郵撥／〇〇〇九九九八——五號
印刷所 三民書局股份有限公司
門市部 復北店／臺北市復興北路三八六號
重南店／臺北市重慶南路一段六十一號
初版 中華民國六十三年七月
增訂十二版 中華民國八十八年八月
編號 S 80014
基本定價 伍元捌角
行政院新聞局登記證局版臺業字第〇二〇〇號
著作權執照臺內著字第一一五四八號

ISBN 957-14-0440-3（平裝）

增訂六版序

本書出版以來，十餘年間，已經數度修訂。內容不斷充實，深獲各界贊許。目前各大專院校，紛紛採作教學之用。工商機構，也多購備參考，或作訓練員工之用。

為了增進工商業與民間的公共關係，此次修訂版特增加了「特種商業書信」一節。內中包括交際函件及其他有關信例，供備參閱。

此外，在金融業務書信和其他部分，也添加了許多新的材料，希望更能適合各界的需要。

程振粵

一九八八年十二月於國立臺灣大學

增訂四版序

本書自面世以來，謬蒙各界人士嘉許，瞬已六版。最近遠在美國紐約的一位讀者來信這樣說：

"坊間所出商用英文一類書籍，迻譯者多不合國情。國人自行編輯者，又多失之簡略。獨尊著內容充實，舉例詳盡，實為最佳之範本……"

對於他的誇獎，雖愧不敢當，但却也給予我無上的鼓勵。

為了適應一般讀者的需要，在這次增訂四版時，除增加各項有關資料外，特將信函部份詳加註釋。對於所舉比較複雜的範例，如信用狀、合同、應用文件等，另譯中文，以利對照閱讀。並增列商業報告一章。

誠懇地希望各界人士批評指教，以便在不斷改進後，能使這本書不但成為一本理想的和完美的大專教材，也是工商界從業人員最好的參考手册。

程振粵

一九八一年三月於臺大

增訂版序

"商用英文"是去年七月出版的。面世以後，因為取材比較新穎，切合實用，很快就受到廣大讀者的歡迎。在短短幾個月之內，初版已經銷售一空。

在本書再版時，除了修正補充之外，特地根據金融界朋友們的意見，加列了"金融業務書信"二十四封，提供銀行從業人員的參考。

目前政府正大力發展我國經濟。對於擴展對外貿易，更是不遺餘力。今後我國國際貿易的範圍愈大，商用英文的應用也愈廣。希望這本書在供大專同學們研習之外， 也能對工商界的從業人員， 有所裨益。

程振粵

一九七五年四月於國立臺灣大學

前　　言

廿多年來，我在臺灣許多國立和私立大學，以及歷屆財政部財稅人員訓練所和經濟部國際貿易專才講習班裏，擔任"商用英文"的課程。最使我困擾的一件事，就是無法找到一本理想的課本。

我認為"商用英文"裏，至少應該包括"商業書信"，"商業電報"，"商業合同"，和"商業文件"四大部份。更重要的，是要多舉與臺灣國際貿易有關的實例，使學者有機會模擬觀摩。

感謝三民書局劉總經理振強，他給了我寫這本書的機會，實現了我多年來的構想。希望對有志學習"商用英文"的人，能够有些幫助。

程振粵

一九七四年七月於國立臺灣大學

商用英文　目次

增訂六版序

增訂四版序

增訂版序

前　　言

第一章　緒　　論

第二章　怎樣寫好商用英文

第三章　商業書信

第四章 商業書信分類實例

第五章　信用狀

第六章　商用電報

第七章　商業合同

第八章 應用商業文件

第九章 商業報告

附 錄

第一章 緒　論

第一節　商用英文的定義

商用英文 (Business English) 是一種爲適應商業需要的標準英文 (Standard English adapted to specific business purposes)。換言之，商業英文並非是另一種英文，只是將普通英文應用在商業方面罷了。正如英文應用在工業、農業、醫學、法學方面一樣，在商業英文裏，會使用不少的商業術語，諸如利息(Interest)、折扣 (Discount)、貨到付款 (C. O. D.)、船邊交貨 (F. A. S.)、信用狀 (Letter of Credit) 等。但就文法和用字造句而論，與普通英文並無不同。所以我們可以說，商用英文就是英文，對於這一點必須有明確的認識。

因此我們要精通商用英文，基本上必須要精通英文。也就是說，英文基礎不好的人，便無法寫好商用英文。但是寫商用英文，和英文作文的寫法却又不一樣。商用英文的要點是簡單、清楚、流利和正確。不需要用複雜的詞句，更要避免使用艱澀的字彙。目的是求收信人能確實明瞭這封信的眞意所在，樂於閱讀。

第二節 為什麼要學商用英文

在答覆"爲什麼要學商用英文"以前，我們應該首先明瞭英文在現代商業上的地位和重要性。自從工業革命以來，英國和美國，一直掌握着世界經濟的樞紐，國際貿易也以英美兩國爲重心。二次大戰以後，美元更成爲各國爭取的對象。因此英文的應用範圍也擴及全世界。可以說英文已是一種溝通全球的媒介。任何人要從事國際貿易，第一個條件便是要精通英文。臺灣的商人和英美的商人來往，固然須用英文，即使和日本、法國、德國乃至南美洲及非洲各國的商人來往，也可以用英文作交易的橋樑。在交易中所用的書信、電報、契約、報告書以及其他商業文件，都以英文爲主。

由於近代交通便捷，商旅往來頻繁，除了文字以外，也增加了各國商人互相面談的機會。因此商業會話也非利用英文不可。商用英文的應用方面，也更趨廣泛。

在競爭劇烈的國際貿易市場上，商用英文已成爲一種克敵制勝的武器。所以有人說，商用英文有"金錢價值"(Dollars-and-cents Value)，因爲商業的目的在於尋求利潤，商用英文可以達到尋求利潤的目的。

再就個人來說，精通商用英文，可以獲得良好工作的機會。各大工商業機構，如銀行、進出口貿易行等，都迫切需要擅長商用英文的職員。這種情形，在臺灣尤其顯著。我們經常在報紙上可以看到高薪徵求這類人才的廣告。

第三節 商用英文的八個條件

怎樣才能寫好商用英文呢？除了良好的英文基礎而外，還需要寫

商用英文的特殊技巧。以下就是衡量商用英文好壞的八個條件:

(1) 清楚(Clearness)

(2) 簡潔(Conciseness)

(3) 正確(Correctness)

(4) 具體(Concreteness)

(5) 愉快(Cheerfulness)

(6) 謙恭(Courtesy)

(7) 體諒(Consideration)

(8) 特性(Character)

先說「清楚」。在寫商用英文時，我們要先問自己，究竟要說的是什麼？確定了所要說的，然後再動筆寫。其次是要使你所寫的，能使對方完全了解，不會發生絲毫的疑問。要避免語意模糊，最好的方法是:

(一)每一句只表達一個意見；

(二)將各句按相關的事項連接起來，使對方易於閱讀及了解。切忌顚倒重複。

(三)用簡單而直接的方式書寫。

什麼是「簡潔」呢？ 就是說在任何商用英文文件裏，所用的文字，必須力求簡潔。每句如此，每段也要如此。電報如此，書信也要如此。例如下面的一封信，我們可以由八十六個字減爲四十二個字。

(86字)

Your letter of September 5 in which you informed us that the goods did not arrive on time, and that three of the desks were damaged beyond repair has been carefully noted.

It is not necessary for us to tell you how much we regret this matter, and we assure you that we shall send you three desks to replace those damaged in shipment.

Thank you for calling this to our attention and hope we may have your further orders in near future, we remain.

Yours truly,

(42字)

In your letter of September 5 we are sorry to learn that our shipment did not arrive on time.

Three desks are being shipped immediately to replace those damaged in transit.

Thank you for calling this matter to our attention.

Yours truly,

此外，每一封商業書信或每一件商用英文文件裏，應該僅限於說明某一件事。這樣不但收件人易於集中精神去閱讀，在案卷方面也便於歸檔和查考。以商業書信而論，每封信的長度以不超過信紙的四分之三地位最爲理想。出口案件固然應當和進口案件分開，即使進口的棉花，也不應和進口的皮革同寫在一封信裏。

但是「簡潔」必須要能表達全部的意思，千萬不可因省略而使信裏的意義不明。

商用英文的第三個條件是「正確」。所謂正確，是指文件的格式要合於規定，英文文法、拼法、標點等不能錯誤。尤其重要的是數目字的正確。如果將 805 元誤爲 850 元，3962包棉花誤爲3762包，或是把收信人的門牌 893 號誤爲 389 號，那麼，便會引起嚴重的後果。因此發信人必須在寄信以前，仔細核對，確保無誤。

再說「具體」，商業文件應避免內容空洞。俗語說得好：「一幅圖畫勝過一千句話」。在商業文件裏，切忌用籠統或過於普通的形容詞。下面便是一個例子：

(籠統的說法)　　　　(具體的說法)

Our asparagus soup is most delicious. Our asparagus soup has a matchless flavor with a richness you'll like better than ever — the richness of luscious, specially grown asparagus blended with fine table butter and delicious seasonings.

比較起來，便知道後一種說法是具體多了。

「愉快」可以帶來友誼、互信、互助、和樂觀。商業書信裏尤其需要愉快的氣氛，使對方閱讀之後，發生好感。在信裏我們應該表示樂意為對方效勞，樂意答覆收信人的一切詢問，貫徹忠誠服務的宗旨。

在商業書信裏，最重要的無過於「禮貌」了。一般說來，信裏總離不了「請」(Please)和「謝」(Thank you)。即使在催收欠款或拒絕要求的信裏，依然要表示充分的禮貌，用極婉轉的語氣。另一禮貌是迅速答覆對方的來信。通常應該是收到的當天就回信。如果是一時不能答覆的事，也應先簡單回信，說明稍遲卽覆的理由。

至於第七個條件「體諒」，是說我們應該在任何商業文件裏，設身處地，為對方的利益着想，切勿主觀。我們在撰稿的時候，必須要問：對方收到這封信或電報以後，會有什麼反應？例如在推銷產品的信裏，我們應該說明顧客買了這種產品之後，可以得到何種好處。卽使在應徵的信裏，也應強調在雇用我以後，對公司有何種幫助。因此「體諒」就是以對方的利益為出發點。這樣就容易得到收信人的好感，達到所希望的目的。

最後談到「特性」。商場如戰場，必須出奇才能制勝。商業書信的寫法，宜運用上面所說的各項技巧，創造獨特的風格。這在推銷新產品，招攬新顧客時，更為重要。

以上的八個條件，因爲英文每字的第一個字母都是“C”，所以又稱爲商用英文的八個“C”。能妥善運用這些條件，寫商用英文便不會發生困難了。

第四節　撰寫人應具備的能力

在明瞭商用英文所需的八個條件以後，接着要知道的便是撰寫商業文件的人，應該具備何種能力，才能達到我們所需要的條件。換句話說，怎樣才是一個撰寫商用英文的高手？下面的六種能力，是他不能缺少的：

(1) 豐富的想像力(A Keen imagination)

(2) 幽默感(A sense of humer)

(3) 了解人性(An understanding of human nature)

(4) 判斷能力(A good judgement)

(5) 精通英文(A good command of English)

(6) 商業道德(Business ethics)

有豐富想像力的人，才能寫出生動的語句，引人入勝。上節所列爲第四條件的「具體」，就非具備豐富想像力的人不能寫出。他能以輕鬆的筆調，描繪成一幅美麗的圖畫，吸引對方的注意。

在商業書信裏，有時需要有幽默的插曲，使對方看了能發生會心的微笑，也符合上節的第五個條件「愉快」。但幽默不可過度，也要因人而異，因地而異。有些民族是喜歡幽默的（例如英國），有些是不欣賞幽默的（例如日本）。在寫信的時候，必須先了解對方的愛惡，以免誤會。

「了解人性」是針對「謙恭」和「體諒」兩個條件的。寫信的人，必須牢記這封信是準備給對方看的，不是只能使自己滿意就可了事。有了解人性能力的人，自然會以謙恭的語氣，處處爲對方設身處地着想，體諒對方的困難。這種婉轉客氣的信，當然會贏得收信人的同情。

判斷能力也是撰寫商業文件必不可少的。既然商場如戰場，判斷的正確與否，和商業的成敗，也息息相關。例如用什麼方法可以有效地推銷新產品？ 用那種動人的文詞可以說服顧客？ 怎樣才能應付競爭？這些都有賴於撰寫人的明智判斷。

至於精通英文，這種能力，更是撰寫人決不可少的。上面所舉的任何條件，都要靠英文來表達。所以我們曾經一再強調，英文不好的人，便無法寫好商用英文。

最後談到商業道德，也就是說，撰寫商業文件的人，要有高尙的品格，誠實不欺。如果信裏所說的和事實完全不符，對方怎能相信？即使被欺，以後再也不會上當了。我們的宣傳，應該根據事實，不能誇張。這樣才能建立商譽，使顧客永遠信賴我們。

第五節　商用英文的範圍

商用英文的範圍很廣。狹義來說，它是指商業書信。誠然商業書信是貿易的最大橋樑。以美國而論，據調查大約有百分之九十左右的交易是經由書信的，僅有百分之十是經由其他途徑。但就廣義來說，商用英文應包括：

(1) 商業書信

(2) 商業電報

(3) 商業契約

(4) 商業報告

(5) 商業廣告

(6) 商業社交文件

(7) 求職和應徵

在商業書信裏，又可分爲下列十一類:

(1) 建立貿易關係(Establishment of trade relations)

(2) 詢問貨價(Enquiry) 和報價(Quotation)

(3) 推銷(Sales)

(4) 定貨(Order)

(5) 交貨(Shipment)

(6) 付款(Payment)

(7) 賠償(Claim)

(8) 保險(Insurance)

(9) 寄售(Consignment)

(10) 求職與應聘(Application for job)

(11) 金融(Banking)

以後各章裏，將列舉以上各類書信的實例，以供觀摩參考。

習 題

(一) 商用英文和普通英文有何不同?

(二) 要具備什麼條件才能寫好商用英文?

(三) 什麼是商用英文裏的八個"C"?

(四) 試述商用英文的範圍。

第二章　怎樣寫好商用英文

在本書第一章裏，我們已經詳細討論過寫商用英文理論方面的各種要素。本章是就實際方面，提示各應行注意事項，分別舉例，加以說明。

第一節　勿用陳腐語句

隨着時代的進步，　商用英文在用詞造句方面，　也有了很多的變更。好些過去在英文書信所習用的字及片語，已不再使用。因此我們在寫信的時候，不宜再用那些陳腐的語句。下面就是這些例子：

(1) Answering, Thanking you 切忌在每句的第一字用「現在分詞」(Present participle)。可改爲 "We answer", "We thank you"。

(2) Advise 有忠告的意思。可改用"tell"或 "inform" 較爲清楚。

(3) And oblige 是不必要的客套語，現已廢棄不用。

(4) As per 應改用 "According to"。例如"As per our telephone conversation" 可改爲 "According to our telephone conversation"

(5) At all times 可用 "always" 來代替。

(6) At this time 可改用 “now”。

(7) Beg to inform you 此種句法現已不用。可改用 “We are pleased to inform you”

(8) By return mail 收到即覆之意，應改用 “immediately”或“at once”

(9) Duly received 意爲收到無誤， duly 一字可省去。

(10) Favor 如係信函， 可直接用 “letter”。如係訂單， 可直接用 “order”。“favor” 一字過於含混。

(11) Has come to hand 可改用 “received”。

(12) In due course, in due time 應說明日期。除事實上無法確定日期外，不能僅說「到相當時期」。

(13) Inst. 意爲「本月」(instant 的縮寫)。 應將月份明白指出，例如本月爲九月，即直接寫明 “September”。

(14) Of the above date 應將日期寫出，不可僅用“above date”。

(15) Our records show 可改用 “We find”。

(16) Permit me to say 是陳腐用語，現已不用。

(17) Proximo 意爲「下月」。應將月份明白指出。例如下月爲十二月，即直接寫爲 “December”。

(18) Recent date 應說明確實日期。 例如 “Your letter of recent date”，可改爲“Your letter of May 5”。

(19) Same 不宜用來代替 “it” 或 “them”。例如 “Your order has been received. We will ship same today”，可改“same”爲“it”。

(20) State 可改用 “say”。例如 “We wish to state” 可改爲 “We wish to say”。

(21) Thank you in advance 預先道謝， 含有對方必須同意的口氣，不太禮貌。

(22) This is to inform you 不切實際，可直接敍述你所要說的事。

(23) Ultimo 指上月。例如 "Your letter of 9th ultimo"，可直接寫爲 "Your letter of July 9"。

(24) Up to this writing 可改用 "till now"。

(25) Valued, esteemed 都嫌恭維過當。"Your valued order"。"Your esteemed firm" 可改爲 "Your order""Your firm"。

(26) We take this opportunity 無意義的開端用語。

(27) With reference to 可改用 "about" 或 "in"。例如 "With reference to your letter of April 4" 可改爲 "In your letter of April 4"。

第二節　愼重修詞

(1) 避免使用不必要的形容詞及副詞

例：Your request is under *very careful* consideration.

The amount is *definitely* correct.

We think the price is *rather* too high.

We enclose *herewith*.

（凡斜體的字，均可省去。）

(2) 避免用不必要的介系詞片語 (Prepositional phrase)

例：We are *in a position* to do it. (*in a position* 可改爲 "able")。

In the course of a few months (*In the course of* 可改爲 "during")。

(3) 避免用無意義的堆砌語句

例：It should be noted that......

We have to point out that......

It will be appreciated that......

這類套語均可删除。

(4) 避免用被動語氣（Passive Voice)

例: "Please be informed" 可改爲 "We are pleased to inform you"。

"He was given a check by us." 可改爲 "We gave him a check."

"Your letter has been received." 可改爲 "We have received your letter."

(5) 避免用過謙語句

例: "Your goodself" 可用 "You" 來替代。

"Only too pleased to" 可改用 "Very glad to"。

"Assuring you of our best attention at all times" 可以省略。

(6) 用簡單的字來代替複雜的字

例: 用 "about" 來代替 "approximately"。

用 "do" 來代替 "accomplish"。

用 "buy" 來代替 "purchase"。

用 "use" 來代替 "utilize"。

(7) 用簡單的字來代替複雜的句子

例: "In the near future" 可改用 "soon"。

"Will you be kind enough" 可改用 "please"。

"For the reason that" 可改用 "because"。

"At the present moment" 可改用 "now"。

(8) 用簡單的句子來代替複雜的句子

例：“It gives me much pleasure to inform you that” 可改爲 “I am pleased to tell you that”。

“We will execute your valued order expeditiously.” 可改用 “We will fill your order soonest.”。

“We do not anticipate any increase in prices in the near future.” 可改爲 “We do not expect prices to rise soon.”。

(9) 在同一句內避免用重複的字

例：“Please quote your *best* price for your *best* quality oil.” 第一個 “best” 可改爲 “lowest” 或 “cheapest”。

“How do you *account* for the fact that the *account* is not correct?” 第一個 account 可改爲 “explain”。

(10) 避免句法單調

誠然，英文商業書信和其他文件所用的句子，應該簡短，使對方容易閱讀。但是如果在一封信裏完全用短句，那麼，又會使讀者感覺單調枯燥。因此能夠酌量配置長、中、短句，便是最理想的了。

例：“We regret that you have not yet delivered the shirts we ordered a month ago.（中句）They are now urgently needed.（短句）The customer for whom we order the goods threatens to cancel his order unless he receives the shirts during this month.（長句）Please deliver them at once.（短句）

第三節 使用正確的標點符號

(1) 句點(Period)(·)

a. 簡寫字的後面都要加句點，這是正常的規定。例如Company的簡寫爲Co.，December的簡寫爲Dec. 等。但有些字是不能簡寫的，例如May, June, July。也有些在簡寫後不用句點的，例如USA本是United States of America的簡寫，可是往往在"U" "S" "A"的後面可以不加句點。這一類的情形很多，如IBM, USIS, CIF, FOB, FAS, FBI, CIA等都是。

有些簡寫的字，基本上已成爲另一個字，所以不能在後面加句點。例如Ad似乎是Advertisement的簡寫，Exam似乎是Examination的簡寫，Confab似乎是Confabulation的簡寫，Gym似乎是Gymnasium的簡寫；實際上Ad, Exam, Confab, Gym都已成爲獨立的字，所以不能在這些字的後面加句點。

b. 在「元」的後面，「角」的前面，應該用句點來分開，這是大家所知道的。但是在整數「元」的後面，不應加句點。例如$4.98是對的。如果是四元，那就不必寫成$4.了。

c. 如果句末的字是簡寫，那麼，就不應在句尾再加句點。例如"He came at 6 p. m."，因爲m. 是簡寫，所以句尾不再加句點。但倘使是疑問句或驚嘆句，那就得在句尾另加疑問號或驚嘆號。例如"Will he come at 6 p. m.?", "Get up, it is already 7 a. m.!"

d. Mr. 和 Mrs. 是 Mister 和 Mistress 的縮寫，但 Miss 不能縮寫。例如 "Miss Wang is our secretary."。

e. 收信人名稱及地址後，信末姓名後，都不用句點。例如

Taiwan Power Company　*　Yours truly,
　　　　　　　　　　　*
Taipei, Taiwan　　　　*　Walter Lee

(2) 逗點(Comma) (,)

a. 在主詞 (subject) 和動詞 (verb) 中間不能用逗點分隔。例如"John, Robert, and Charles are my good friends." Charles 和 are 中間不能用逗點分開。

b. 如果句中最後兩個名詞是作爲一個單位，那麼，在 and 之前，不能用逗點。例如 "I eat meat, egg, bread and butter every day." 因爲 bread and butter 作爲一個單位，所以 bread 後不需要逗點。

c. 在 Jr. Sr. 和 Esq. 等稱謂後面，應用逗點。例如："Mr. Charles, Jr., is our customer"; "Walter King, Esq., is not in town."

d. 在引用句 "Quotation" 前，應該用逗點。例如 The teacher said, "I must go home now."

e. 用逗點來分開地名和日期。例如 "186 Nanking Road, Taipei, Taiwan" 和 "March 20, 1976"。

f. 用逗點來分隔數目。例如"$1, 572, 369"; "56, 740 bales of cotton"。 但年代、電話號碼、門牌號碼、頁數、保險單號碼等，不能用逗點來分開。例如 "1976"; "1275 Aster Road"; "Page 5431"; "Policy 742166"。

(3) 冒號(Colon) (:)

a. 在商業信函的稱謂 (Salutation) 後，都用冒號。例如

"Dear Sirs:"; "Gentlemen:" 私人函件，可用逗點。例如"Dear Jane,"; "Dear Mother,"。

b. 用冒號來分隔時和分。例如 "4:15 p. m.";"6:40 a. m."。

c. 用冒號來引述下列各項。例如 "The order included the following items: 5 pencils, 6 books, and 2 maps."。

(4) 引用號(Quotation marks) ("　")

a. 在強調一個名詞或一件事時，可用引用號。例如 "The term "super" is used freely today."; "The "training program" must be carried out right away."。

b. 如果在引用句裏再有引用名詞，引用名詞可用單引用號。例如 The teacher asked, "What is meant by the term 'shorthand'?"。

(5) 括弧號(Parentheses) (( ))

a. 爲確保數目無誤，商業文件內常在文字後再用數目字放在括弧號內。例如"We received sixty (60) orders this week."; "The amount is three thousand and sixty-five dollars ($ 3, 065)."。

b. 作指示或參考之用。例如"The Second World War was first fought in France (see page 5 of The Modern History)"。

(6) 所有格號(Apostrophe) (')

a. 普通只能用在有生命的東西上。例如"Daisy's hat"。對於無生命的東西，應用 of。例如"The leg of the table", 不是 "The table's leg"。但時間、衡量、人化等，又可同樣作爲有生命的東西。例如 "A day's work"; "A mile's length"; "For pity's sake."。

b. 在公司或社團名稱裏，常將所有格號省略。例如 “The Farmers Bank of China”; “Workmans Cooperative Society”

(7) 聯字號(Hyphen) (-)

a. 商業文件打字到每行的終了時，如果最後一字比較長，爲了不使那一行突出，可將這個字分開，將後一部份打在次行。但聯字號必須放在這個字前半部的後面，絕對不能放在後半部的前面。換句話說，每行的第一字前面，不能有聯字號。例如信的某行最後一字是 “com-merce”，如果要分做兩行，那麼，第一行是 “com-”，第二行開始是 “merce”。

b. 將數目和名詞連接，成爲形容詞。例如 “A 50-mile race”; “An eight-hour day”。名詞不能用複數。

c. 有些字是由兩個字連接而成，中間有固定的聯字號，不能省略。例如 “first-rate”; “self-explanatory” 。但也有些字雖係兩字合成，却不能在中間加聯字號的。例如 “friendship”; “herself”; “whenever”; “childhood”。

第四節　大寫字體的用法

(1) 每一句的第一字固然應該大寫，即使在引用句裏，也要如此。例如 The boy said, “My father is 76.”。

(2) 月份、週日，節日都要大寫，但春、夏、秋、冬不必大寫。例如“September”, “Monday”; “Christmas”; “I shall spend the winter in Japan.”。

(3) 專門名詞、人名、地名都應大寫。如果已作普通名詞用時，就不必大寫。例如 Diesel 是德國人，當然要大寫，但 diesel engine（柴油機）已是普通名詞，所以 diesel 不必大寫。同樣的情形，有 "roman candle", "china cup" 等。

(4) 方向不必大寫，例如 "The west wind"; "We shall drive north"。但如果是指示一定的方位時，就應大寫。例如 "The war in the Middle East"; "He came from the North."

(5) 句內的冠詞 (article) 不必大寫，但如係公司規定名稱的一部份，就應大寫。例如 "We went to The First Commercial Bank of Taiwan.";"We wrote to the Taiwan Power Company."。

(6) 有牌名的產品，雖然產品是普通名詞，仍應全部大寫。例如 "Underwood Portable Typewriter"; "Tatung Radio."。

(7) 某某公司內部各部門的名稱，均應大寫。例如 "Sales Department of IBM"; "Personnel Department of SONY Co."。但一般性的，不用大寫。例如 "He works in the personnel department of one of our banks."。

(8) 在姓名前的尊稱應大寫。例如"I saw President Gerald Ford." 如用尊稱來代替某人的，也應大寫。例如 "I wrote to Mother."; "I called Father yesterday."。但後者如在尊稱之前有冠詞或用所有格的，尊稱即不用大寫。例如 "I called my mother."; "He may be our next president."。

(9) 商業信扎中的開端稱謂 Dear Sirs 二字均應大寫。信尾謙稱 Yours truly，只須將 Yours 一字大寫。

(10) 文件中的某部份，如 "Section 3"; "Part V"; "Article 14" 等均應大寫。但頁數不必大寫，如 "page 35"。

(11) 在重要商業文件，如合約之類，內中錢數，都須大寫，以昭愼重。例如"The rent is Three Thousand and Four Hundred Dollars each month."。

(12) 學校中的學科名稱，如 economics, accounting, public finance, statistics 等等，無須大寫。各級大學生的 freshman, sophomore, junior, senior 也不用大寫。

第五節　數目字的寫法

(1) 重要文件以外，一般商業書信裏都是用數目字來表示錢數，例如 $ 81. 75 或整數 $ 62。但如果在同一句內，有幾個不同的數目，或整或零，那麼，爲了劃一起見，整數後面，也必須加零。例如"We paid the following bills yesterday: water, $ 16. 54; telephone, $ 75. 00; gas, $ 30. 50; and electricity, $ 130. 00"。

(2) 在重要商業文件裏，除了將錢數全部用英文寫出外，也有再在後面註明數目字的。例如 "The price is Seven Hundred Dollars and Sixty Cents ($ 700. 60)."。

(3) 百萬以上的整數，爲了避免用太多的圈，可用 million, billion 等字來代替。例如 "We sold three million TV sets."，不必用 "We sold 3, 000, 000 TV sets."。

(4) 句子第一個字不能用數目字。例如 "Ninety students are here."，不能寫成 "90 students are here."。

(5) 日期應用數目字。如 January 1, 1977, 不能用 January first, 1977。但在正式文件裏，却要寫出，甚至年代也要全部寫出。例如 "This Agreement was made on the first day of April in the year of one thousand nine hundred and seventy

seven."。

(6) 街道號碼和門牌號碼如果用在一起，那麼，門牌號碼用數目字，街道號碼却要用英文寫出。例如 "85 Fifth Avenue"。

(7) 百分數應寫爲 "per cent"，二字分開，不要連爲一字，更不要用%符號來代替。例如 "Our salary has increased 10 per cent last year." 但在利息和折扣方面，可以用%。例如"Our price is subject to 2% discount if paid within 60 days."; "The interest of this bond is 5%."。

(8) 上午的時間用 a. m. 下午用 p. m.，如果一句內有兩個時間，一爲整數，一有分數，那麼，爲劃一起見，整點也要加圈。例如"The bank will open at 9:00 a. m. and close at 5:15 p. m."。倘只說 "The bank will open at 9 a. m.";那 9 字以後即無須加兩個圈了。

(9) 如果兩個數目連在一起，必須用逗點分開，以免誤會。例如 "On September 12, 1977, 19, 625 TV sets were sold."

(10) 書報的頁數，一律用數目字。例如 "Please see page 118 of this book."。

(11) 度量衡通常都用數目字。例如 "We ordered 150 bales of cotton."; "The room is 40 by 20 feet."; "The temperature reached 95 degrees yesterday."。

(12) 號數 (Number) 通常用No. 或 # 來代表，例如 "Your policy No. 5721 was canceled." 不能寫成 "Your policy Number 5721 was canceled."。

習 題

試將下面的信使用正確的標點，並將應該大寫的字，分別大寫。

mr john f donahue 1148 klondike street nome alaska dear mr donahue did you know that termites in this country have an annual "grocery bill" of over $50,000,000 and that you without suspecting it may be paying your share to support these wood-eating pests (*new paragraph*) termites are tiny insects that nest in the ground and come up to attack the wood in buildings once inside the wood they work back and forth eating away until finally costly damage has occurred(*new paragraph*) termites always work within the wood rarely coming through the surface consequently it is almost impossible for the homeowner to detect their presence before serious damage has been done (*new paragraph*) the only way to be sure that termites are not damaging *your* home is to have it inspected by a termite expert our nearest licensee will be glad to send out a trained termite inspector to examine your property for termites at your request

第三章　商業書信

第一節　格　式

商業書信的格式，主要的有三種，一種是閉塞式(Block Form)，另一種是半閉塞式(Semi-block Form)，還有一種是改良閉塞式(Modified Block Form)。

閉塞式（圖一）是信裏的所有文字，包括日期和簽名在內，全部從信的左邊開始。這個格式現在很流行，因爲打字員在打信時，可以不用計算左邊應留的地位，比較省事。但是它的缺點是整個的信，傾向左邊，顯得異常不平衡，有欠美觀。

半閉塞式（圖二）是將日期放在信的右邊，簽名放在中間偏右，這樣就顯得兩邊平衡。此外在每段開始時，向裏縮入五個字母的地位。改良閉塞式（圖三）和半閉塞式相似，所不同的是每段開始時緊靠着左邊，不縮入五個字母的地位。

以上三種格式，都可採用，一般說來，採用半閉塞式的比較多。至於收信人及地址後面的標點，現在的商業書信裏一概不用。

除了圖一、二、三的三種格式以外，現在美國也流行一種“NOMA簡化式”(NOMA Simplified Letter)。這種新型格式，是由美國國家

圖一 閉塞式

第一商業銀行
FIRST COMMERCIAL BANK
15 CHUNGKING SOUTH ROAD SECTION 1, TAIPEI
TAIWAN, R. O. C.

CABLE ADDRESS: "FIRSTBANK" TAIPEI
TELEX: 11310, 11729
11740, 11741

P. O. BOX 395 TAIPEI

March 1, 1989

Mr. John Stewart
57 Min Chuan Road
Taipei

Dear Mr. Stewart:

First Commercial Bank wishes to extend to you a cordial invitation to open an account with it and to make use of the comprehensive banking facilities it offers. Our bank is conveniently located near your office, and our experienced staff will always be glad to render you prompt and efficient service.

As you are already aware, First Commercial Bank with its worldwide network of branches and correspondents, is in a position to offer you every modern banking facility.

It is sincerely hoped that we shall have your patronage in the very near future.

Very truly yours,

K. S. Wang
Manager

KSW:ac

事務管理協會 (National Office Management Association) 所倡導的（圖四）。特點是 (1) 和閉塞式相同，信裏的所有文字，包括日期和簽名等在內， 全部從信的左邊開始。 (2) 每信必須有「事由行」(Subject line)，並且要完全大寫。(3) 信末發信人的姓名也要大寫。(4) 收信人的姓， 在信的第一段第一句裏要提到。(5) 每行的最後一字，不論有多長，不能因爲顧全信的兩邊平衡而予以分割。現在美國有許多大公司採用這種信的格式，一方面由於簡便，一方面也是因爲

圖二　半閉塞式

第一商業銀行
FIRST COMMERCIAL BANK
15 CHUNGKING SOUTH ROAD SECTION 1, TAIPEI
TAIWAN, R. O. C.

CABLE ADDRESS:
"FIRSTBANK" TAIPEI
TELEX: 11310, 11729
11740, 11741

P. O. BOX 395 TAIPEI

March 1, 1989

Mr. John Stewart
57 Min Chuan Road
Taipei

Dear Mr. Stewart:

First Commercial Bank wishes to extend to you a cordial invitation to open an account with it and to make use of the comprehensive banking facilities it offers. Our bank is conveniently located near your office and our experienced staff will always be glad to render you prompt and efficient service.

As you are already aware, First Commercial Bank with its worldwide network of branches and correspondents, is in a position to offer you every modern banking facility.

It is sincerely hoped that we shall have your patronage in the very near future.

Very truly yours,

K. S. Wang
Manager

KSW:ac

職員中有不少是國家事務管理協會的會員。但是這種格式，並非普遍適用，尤其在國際貿易方面，更不相宜。

如果以同樣的信，分寄給很多人，那麼，可採用"通函式"(Displayed Letter)（圖五）。這種格式， 在信上沒有收信人的姓名和地址，也沒有稱呼。認爲重要的部份，可以全部用大寫。也可以用不同的顏色或在下面劃線來強調重要的語句。但強調的地方不宜太多，以免影響整個信的重要性。

圖三 改良閉塞式

第一商業銀行

FIRST COMMERCIAL BANK

15 CHUNGKING SOUTH ROAD SECTION 1, TAIPEI

TAIWAN, R. O. C.

CABLE ADDRESS:
"FIRSTBANK" TAIPEI
TELEX: 11310, 11729
11740, 11741

P. O. BOX 395 TAIPEI

March 1, 1989

Mr. John Stewart
57 Min Chuan Road
Taipei

Dear Mr. Stewart:

First Commercial Bank wishes to extend to you a cordial invitation to open an account with it to make use of the comprehensive banking facilities it offers. Our bank is conveniently located near your office, and our experienced staff will always be glad to render you prompt and efficient service.

As you are already aware, First Commercial Bank with its worldwide network of branches and correspondents, is in a position to offer you every modern banking facility.

It is sincerely hoped that we shall have your patronage in the very near future.

Very truly yours,

K. S. Wang
Manager

KSW:ac

通函式的信， 格式比較自由。 同時也是用複印方式根據名單分發，收信人的姓名住址，僅在信封上可以看到。

爲了簡化商業書信，有些公司採用明信片來通知或詢問顧客簡單事項。上面不列收信人的姓名住址，也沒有信尾謙稱。這種明信片是附在公司致顧客的信封裏，或夾在刊物裏，由顧客在明信片上塡好後付郵，不用貼郵票。這是便利顧客答覆的一種簡捷辦法。稱爲“私用郵寄明信片”(Private-mailing Card)（圖六）。

圖四　NOMA 簡化式

NASHUA CORPORATION
NASHUA, NEW HAMPSHIRE

March 19, 1989

Mr. Charles Chang
1109 East Hollywood Avenue
Los Angeles, California

OUR AGREEMENTS TO THE FORTUNE MAGAZINE

Your letter of March 1, 1981, Mr. Chang, made us to doubt whether you know the two points on which we are committed to take action. The points are:

1. We agree to advertise in the Fortune Magazine's new edition each time it is issued.
2. We are going to pay the fees every three issues at a special rate of 65%.

Please let us know when our advertisement will appear in the new edition of the Fortune Magazine.

RICHARD S. KING

ec3

銀行和規模比較大的企業，如電信局、電力公司、煤氣公司、自來水公司等，爲簡化起見，常採用“格式信”(Form Letter) 來處理例

圖五 通函式

The Business Club. 148 Main Street, Dallas, Texas

June 10, 1989

TO ALL MEMBERS OF THE BUSINESS CLUB

This year, for the second time, The Business Club will again have an opportunity to participate in the International Business Exhibition. This exhibition will be held in Los Angeles for one month, beginning September 4.

NOW IS THE TIME TO DECIDE WHAT YOU CAN DO TO MAKE OUR PARTICIPATION A SUCCESS. EACH MEMBER SHOULD BEGIN AT ONCE TO MAKE PREPARATIONS FOR THIS IMPORTANT EVENT.

Our time is limited. Each one of us must act quickly if we are to make the best possible use of this great chance to publicize the many activities of THE BUSINESS CLUB.

PLEASE DO YOUR BEST!

Sincerely,

Daisy S. C. Chen, Secretary

行公文。這種是印好的格式，只須塡上收信人姓名住址、日期、和其他有關事項卽可。寄發時採用「窗式信封」（圖十四），所以信封上不必再打收信人的姓名及住址了。（圖七）

「格式信」只能應用於極簡單而例行的事件。優點是可以迅速處理大宗函件。但缺點也很多。最顯明的是收信人看到這種信，往往缺

圖六　（甲）私郵明信片（正面）

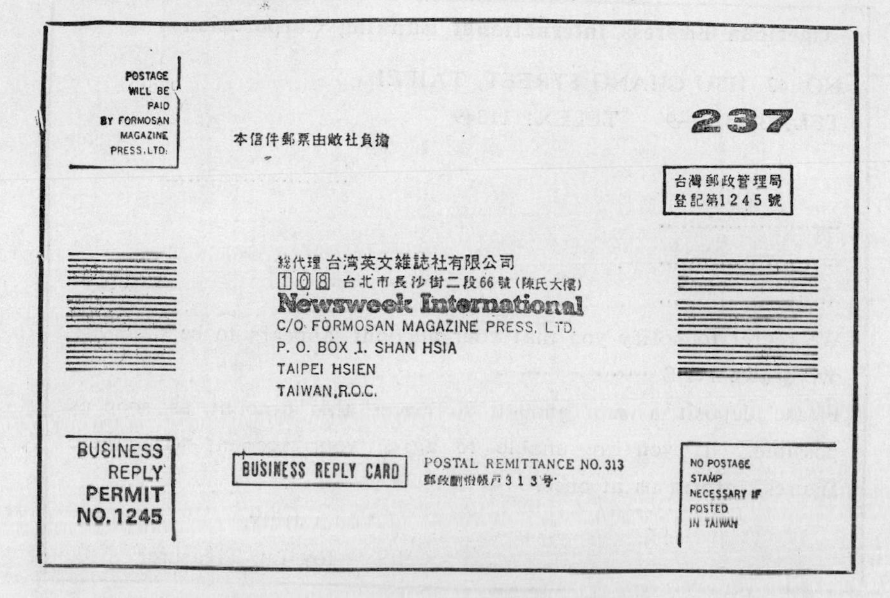

POSTAGE
WILL BE
PAID
BY FORMOSAN
MAGAZINE
PRESS.LTD.

本信件郵票由敝社負擔

237

台灣郵政管理局
登記第1245號

總代理 台灣英文雜誌社有限公司
108 台北市長沙街二段66號（陳氏大樓）
Newsweek International
C/O FORMOSAN MAGAZINE PRESS, LTD.
P. O. BOX 1. SHAN HSIA
TAIPEI HSIEN
TAIWAN,R.O.C.

BUSINESS
REPLY
PERMIT
NO.1245

BUSINESS REPLY CARD

POSTAL REMITTANCE NO. 313
郵政劃撥帳戶３１３号

NO POSTAGE
STAMP
NECESSARY IF
POSTED
IN TAIWAN

圖六　（乙）私郵明信片（反面）

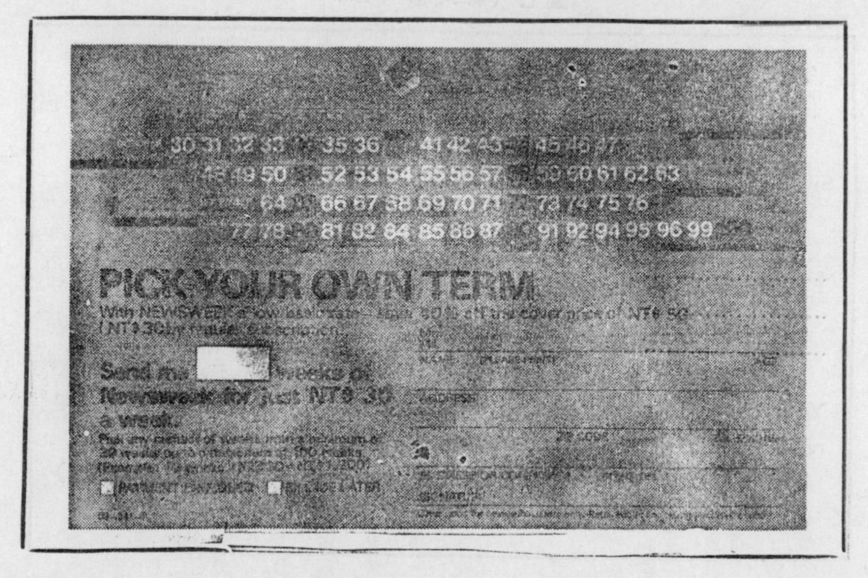

圖七　格式信（甲）

American Express International Banking Corporation

NO. 42, HSU CHANG STREET, TAIPEI
TEL: 3715271-9　　TELEX: 11349

....

....
....
....

We regret to notify you that your account appears to be overdrawn NT $ ....

Please deposit a sum enough to cover this amount as soon as possible. If you are unable to agree your account with these figures, inform us at once.

Yours truly,
Joy Lu, Treasurer

圖七　格式信（乙）

THE AMERICAN TELEPHONE & TELEGRAPH COMPANY

Cashier's Office
283 State Street, Boston 24, Mass. ,

....

....
....
....

Your check in connection with your account is returned for the reason indicated below, marked (x).

Signature omitted ( )　　Made out to other than this **Co.** ( )
Endorsement omitted ( )　　Dated ahead ( )

乏親切感，不予重視。爲了補救這個缺點，可以採用下列五種方法:

(1) 用好的紙張。

(2) 除日期及收信人姓名地址用打字機打出外，稱呼和信尾謙稱，也應打字，不必預先印好。

(3) 除非信件太多，最好每封信上親自簽字。

(4) 信封必須封妥。

(5) 不要作印刷品郵遞。

大公司爲了節省人力，往往準備一種“格式分段書信”(Form Paragraph Letter) 手册。這個手册，內中按業務性質分爲若干類，例如放款類、催收帳款類、推銷類、賠償交涉類等等。每類給予一個代號 (Code number)，下面由撰稿人就各種可能發生的不同情況寫成若干段，每段也編號。例如放款類下面有五段，那麼，編號就是 $L_1L_2L_3L_4L_5$。假定銀行收到一封要求貸款的信，經理在回信時，只消按答覆的意思，在一張便條紙上註明手册上的編號，例如 $L_1L_3L_5$，交給打字員。打字員就立即照手册上所規定的三段，作爲覆信的信文，加上日期，收信人姓名及地址、稱呼、信尾謙稱，發信人姓名等，打字後送請經理簽字寄發。這種方式的回信，有下列各項優點:

(1) 手册內的各段信文，已由高手擬妥，不須另行撰稿。

(2) 全信用打字，和一封普通的答覆函，沒有兩樣。換句話說，依然富於親切感。

(3) 打字員照手册上的信文打字，不易發生因原稿模糊而造成的錯誤。

當然，這種「格式分段書信」，不僅可供答覆來信之用，公司主動發出的信，也可適用。例如催收款項時，就可酌用「催收帳款類」的若干段（假定是 $C_1C_4C_6C_9$）。

「格式分段書信」既然要編成手册，又要請高手就各種不同的情況分類分段撰稿，因此也只有大規模以及業務複雜的企業才宜應用。

第二節　商業書信的九個部份

商業書信的構成，可分爲九個部份：

(1) 信頭(Letterhead)。

(2) 日期(Date line)。

(3) 收信人名稱住址(Inside address)。

(4) 稱呼(Salutation)。

(5) 信文(Body of the letter)。

(6) 結尾謙稱(Complimentary close)。

(7) 簽字(Signature)。

(8) 發信人及打字員姓名第一個字母(Identification initials)。

(9) 其他(Others)。

第一項　信頭

信頭是指信紙上面印好的公司行號名稱、地址以及其他事項。信頭的設計要美觀，但以簡單爲宜。有些除名稱地址外，還列有電話及電報掛號、公司的主要業務、主管人的姓名等。倘使有商標的，也可以將商標印上。（圖八）

如果不用印有信頭的信紙，那麼，寄信人的名稱及地址，應該打在信紙上的右上角，在日期上面。（圖九）

第二項　日期

商業書信上的日期，打在信頭下面四行至六行的右方。月份不能

圖八　信頭的各種設計

RELIANCE Products Corporation

an amtel company

148 Lime Street, London, W.C.4, England

SCIENTIFIC INTERNATIONAL INC.
P.O. BOX 1590
BLOOMFIELD, N.J. 07003 U.S.A.
(201) 743-1892

CABLE: SCIENTIFIC

Charles Kempster
Furniture Repairer and Restorer
The Old Forge Broad Lane Richmond Surrey RI6 3BZ
01-823 7325

T. J. HOLMES CO., Inc.
MANUFACTURERS OF ATOMIZER
Chartley, Mass.

第一商業銀行

FIRST COMMERCIAL BANK
15 CHUNGKING SOUTH ROAD SECTION 1, TAIP
TAIWAN, R. O. C.

CABLE ADDRESS:
"FIRSTBANK" TAIPEI

TELEX: 11310, 11729
11740, 11741

(a)

用數字來代表，必須完全用英文字寫出。因為用數字代表月份的用法，英美兩國不同。例如 7. 2. 1981 在英國是指一九八一年二月七日，在美國則指一九八一年七月二日。

日期有下列幾種寫法：

(1) July 2nd, 1981

(b)

(2) 2nd July, 1981

(3) July 2, 1981

雖然三種都可採用，但最普通而又最明顯的，還是第三種。也是現在英文書信上所常用的。

圖九 沒有印信頭的信

Hwa Hsin Commercial Bank
56 Nanking East Road, Sec. 2
Taipei, Taiwan
September 1, 1989

Mr. Johnson S. Chiang
239 Ta Tung Road
Taipei, Taiwan

Dear Mr. Chiang:

We regret to inform you that we are not in a position to grant you a loan for NT$50,000 as requested in your application dated August 16, 1981.

As our sources of loanable funds are quite limited, our loan policy is set to extending loans only to those borrowers who can give us the most collateral benefit. This means that our loan borrower should give us business at least for six months before we can consider making a loan to him.

Again, we, as a policy, are only making short-term loans up to six months which must be paid upon maturity. Any loan in excess of a six-month period is beyond our consideration.

Although we are not in a position to assist you this time, we suggest that you make use of our other facilities for which we shall be pleased to serve you.

Very truly yours,
William K. Lim
Manager

WKL:scd

第三項 收信人名稱住址

收信人的名稱和地址，應該低於日期二行至四行，打在信紙的左邊（圖九）。但也有放在下面左邊，在簽字下二行至四行的。

收信人如果是公司行號，那麼，應該用它的全名。有些公司前面有 "The" 字，有些沒有。例如臺灣銀行的英文名稱為 "Bank of Taiwan"，前無冠詞。但是以前的臺灣第一商業銀行，英文却是 "The First Commercial Bank of Taiwan"。有些公司用 "Company"，有的用 "Corporation"，例如臺灣電力公司的英文名稱是 "Taiwan Power Company"，中國石油公司却是 "China Petroleum Corporation" 前面都沒有冠詞。我們寫信給它們，決不可任意變更它們原定的名稱。美國鋼鐵公司是"The United States Steel Corporation"，但通用汽車公司則為 "General Motors Corporation"。"The United States Steel Corporation"，也不能簡寫為"The U. S. Steel Corporation"。

如果商業書信是寫給公司經理的，應在公司名稱上面加 "The Manager" 例如:

The Manager
Bank of Taiwan

倘使是寫給公司裏的個人的，應該把他的姓名寫在公司名稱的上面。例如:

Mr. James Wang
Martins Bank, Ltd.

收信人地址緊跟在公司的名稱下面，一般分為三行。例如:

78 Baker Street
San Jose, California 95119
U. S. A.

英國人習慣在門牌後加 ","，美國人則簡略不用。

如果地址有巷有弄的，街道又分段的，那麼，應照下面的寫法:

5 Alley 32, Lane 15, Nanking East Road, Sec. 4
Taipei, Taiwan
Republic of China
(中華民國臺灣臺北市南京東路四段十五巷三十二弄五號)

第四項 稱 呼

寫給公司行號的信，在開頭的稱呼，普通有兩種：

(1) Dear Sirs:

(2) Gentlemen:

雖然英國人習慣上比較喜歡用"Dear Sirs"，美國人多數用"Gentlemen"，事實上兩者都可隨意使用，並無任何限制。

如果收信人是"The Manager"，而無姓名，稱呼只可用 Sir 的單數，也就是"Dear Sir"。但不能用"Gentlemen"的單數"Gentleman"。同樣的，在商業書信裏，"Sir"不能單獨使用，必須在前面有"Dear"。

假使收信人是有姓名的，那麼，便要在稱呼時將姓寫出。例如：

Dear Mr. Wang:

一般規定是商業書信在"Dear Sirs"後面應該用"："。私人信函，在稱呼後面用"，"。例如"Dear Della,""Dear Mother,"之類。

第五項 信 文

信的內容，也就是信的主體。這是全信最重要的部份，視信的性質，分爲若干段。半閉塞式在每段開始時，縮入五個字母的地位。兩段之間，要空一行。

信文應在稱呼下面兩行開始，用單行 (Single spacing) 最相宜。如果因爲信短而必須用雙行 (Double spacing)，未始不可。但兩段之間，仍然只能空一行，這是要特別注意的。

商業書信內容，整個約佔用信紙的四分之三位置，兩邊及上下各留少許空位。空位的大小，須配合信的長短。如是長信，左右兩邊至少要空出一英寸的地位。上下的空位，也要相等。印有信頭的信紙，上下約空一英寸半。如用第二頁，那麼，在白紙上端，也應空一英寸半。倘是短信，除兩邊所留地位不變外，上下的空位，可以酌量放大。每段文字不可太長，尤其是開始和末尾兩段，更應簡短。

如用閉塞式，每段應從信的左邊開始。如用半閉塞式，每段開始應自左邊縮入五個字母的地位，也就是從第六個字母起打字。

第六項　信尾謙稱

信尾謙稱，在商業書信裏，也有好幾種:

(1) Yours truly,

(2) Truly yours,

(3) Yours very truly,

(4) Very truly yours,

(5) Yours faithfully,

(6) Faithfully yours,

倘收信人不是公司行號，而是個人，那麼用"Yours sincerely,"或'Sincerely yours,"也可以。在信尾謙稱的後面，都是用“,”。照半閉塞式，地位應該在信文最後一段下面兩行的中間偏右邊。

第七項　簽　字

因爲簽字往往難於辨認，所以在信尾謙稱下面要空五行，將發信人的姓名清楚打出。發信人就可利用這五行的空位簽字。

在商業信扎裏，發信人的姓名下，應該將他的職銜也打出。下面的三個樣子，自然以乙式爲最合宜。

Vice President In Charge of Operations	J. A. MIREL Treasurer	
(甲式) 沒有將姓名打出，簽字看不淸楚。	(乙式) 標準格式	(丙式) 沒有將姓名打出，也沒有職銜。

倘使簽名的是女人，必須在打字的姓名前加註 (Miss) 或 (Mrs.)。舉例如下:

Sincerely yours,	Yours truly,
(Miss) J. E. Anderson	(Mrs.) Alice Brooks

如果對方來信沒有註明(Miss)或(Mrs.)，我們回信時可用(Ms.)，這樣對小姐及太太都可適用。

第八項 發信人及打字員姓名的第一個字母

爲了便利查考，商業信扎裏常將發信人及打字員姓名的第一個字母 (Initials)，打在簽字地位下面兩行的左邊。(參閱圖一至圖三) 但也有幾種不同的方式:

(1) 發信人的首字用大寫，打字員的用小寫。例如:

CYC:dc　　CYC/dc　　CYC dc

(2) 兩人的姓名首字都用大寫。例如:

CSM:FT　　CSM/FT　　CSM FT

現在一般使用的是下面兩種方式。例如:

SWK:ml　　CKM:GL

第九項　其　他

除了上面八種，有時候還可以增加下列其他項目：

(1) **附件**　如果信裏說明有附件，應該在信的左邊最下面，也就是在發信人及打字員姓名首字的下面，寫明 Encl.。倘使是一張支票，就寫 Encl.: One check。爲了簡化起見，常在 Encl.: 後面寫 "As stated" 或 Encl.: a/s。中文的意義，便是"附件如文"。

(2) **附言**　有時信已簽字，但臨時要加幾句話，可在信的最下面一行，寫P. S. 兩個字母，也就是Postscript（附言或再啓者）的縮寫。除非確屬必要，最好不要用"附言"。P. S. 後不要加 "："。

有時 P. S. 也用來強調信裏所說的某一件事。例如：

P. S.　To save time please wire us the result.

P. S.　Mr. Cheng is scheduled to leave for New York via CAL tomorrow.

(3) **增加頁數**　理想的商業英文書信，是打字的部份，約佔全頁信紙的四分之三。倘使是一封比較長的信，非增加第二頁不可，那麼，第二頁的最上一行，應該將收信人的姓名寫在左邊，中間寫 "2" 或 "Page 2"。右邊打日期。例如：

	Mr. John Stewart	2	March 1, 1981
（或）	Mr. John Stewart	Page 2	March 1, 1981
（或）	Mr. John Stewart	—2—	March 1. 1981

第二頁應該用沒有信頭的白紙，切不可再用印有信頭的信紙。紙質也要一樣。

第二頁上至少要有信文三四行。切忌只在第二頁打信尾謙稱及簽字，這樣會使信的形式極不調和。

(4) **加"注意"或"事由"** 有時商業書信，在稱呼的同一行，加"請某人注意"一行。或加"事由"一行。舉例如下：

1.

China Petroleum Corporation
98 Yen Ping Road
Taipei

Dear Sirs: Attention: Mr. Charles Lee

2

Bank of Taiwan
Chungking South Road
Taipei

Gentlemen: Subject: Your enquiry of October 15

將事由放在稱呼下，正文前也可以。如果"請某人注意"和"事由"兩項都用，就必須將"請某人注意"放在稱呼同一行，將事由放在稱呼下，信文上兩行。例如：

The Richard Corporation
123 Nanking East Road, Sec. 1
Taipei

Gentlemen: Attention: Mr. James Chen

Subject: Your Order No. 1189 of February 1, 1989

(5) **副本抄致** 將信函副本，抄送有關機構或個人，應在信的左下端打"C.C."(Carbon Copy 的簡寫)字樣。例如：

C. C. Bank of Taiwan, Taipei (或)

C. C. Prof. C. Y. Cheng, National Taiwan University

如須抄送幾個單位時，只可打一個"C. C." 例如:

C. C. Mr. Walter S. Wang, 301 Nanking Rd., Taipei
Dr. Y. S. Ma, 47 Fifth Avenue, New York

(6) 郵遞區號 為加速揀信及遞送起見，郵局特定了一種"郵遞區號"制度，英文稱為 ZIP Code。以美國而言，區號共計五個數字。首三個數字代表收信人地址的區域，後面兩個數字代表收信人的分區。郵遞區號應該打在州名後面三個字母的地位，在同一行上。舉例如下:

International Trading Company
1145 Park Avenue
New York, N. Y. 10022
U. S. A.

各國所定的郵遞區號，並不相同。加拿大的却很特別。數字之外，又摻雜字母，例如:

Dr. Y. C. Ma
34 Acadia Bay
Winnepeg, Manitoba R3T 3H9
Canada

臺灣的郵遞區號，只用三個數字，因地區狹小的緣故。許多國家，還沒有採用這種制度，或僅採一種簡單的分區代號。例如倫敦的 N. W. 2，就是代表倫敦西北第二區。一九八八年臺灣也改郵遞區號為五個數字，正普遍推行中。

第三節 使用的文具

第一個印象是最令人難忘的。商業書信的目的，在取得對方的好感。因此，信的外表和信的內容，同樣重要。有時甚至於外表比內容更重要。因為收信人在閱讀信的內容以前，一定注意到信的外表。在外表方面，有下列各點要特別注意：

(1) 信紙要白色，沒有橫格的；品質也要最好的。

(2) 信頭的設計要美觀，但不要太複雜。

(3) 普通的信紙，應爲八英寸半濶，十一英寸長。這種尺寸，比較歸卷方便。也有用七英寸又四分之一濶，十英寸半長的。作備忘錄用的短信紙，通常爲五英寸半濶，八英寸半長。

(4) 打字要正確無誤，不可塗改。

(5) 打字帶色澤要濃淡適中，清新悅目。

(6) 上下左右的空位要均勻。

第四節　信封的寫法

寫信封的時候，要把收信人的姓名和地址放在信封的下半部略靠右方的地方。把地址的每一單項，寫在單獨的一行。例如第一行爲收信人姓名（或公司名稱），第二行爲街名及門牌，第三行爲城名及州名，第四行爲國名。（國內通信，不寫國名。）

信封上的收信人姓名和地址，必須正確無誤，以免誤投。美國郵局統計每年有三千萬至三千五百萬封信，因地址或收信人姓名錯誤而無法投遞。如果收信人及地址不超過四行，應在每行之間多空一行，也就是 “Double Spacing”，使郵局容易辨認。（圖十）

如果我們將信封分爲四個部份（圖十一），那麼，記在 “A” 位置上的是發信人的行名地址。在它上面，也有加印：

(1) If undelivered, please return to
（無法投遞，退還原處。）或

(2) After 10 days, please return to
（十天後請退回原處。）

圖十　信封格式

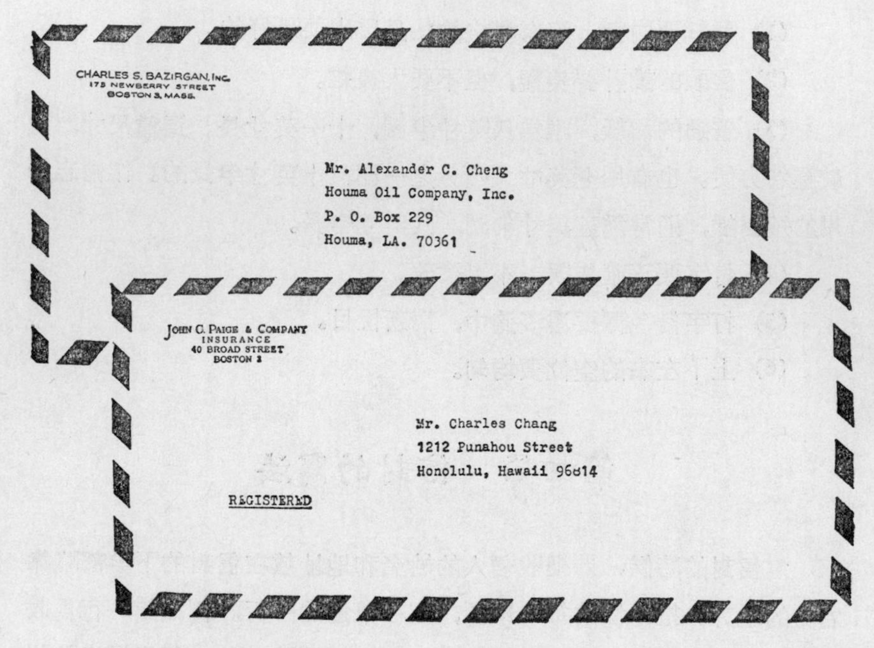

記在"B"位置上的是:

(1) 印刷品(PRINTED MATTER)。

(2) 樣品(SAMPLE)。

(3) 無價值的樣品(Sample of No Value)。

(4) 贈品(With Compliments)。

(5) 商業文件(Commercial Paper)。

貼在"C"位置上的是郵票。郵票最好祇貼一枚，例如寄美國的航空信，貼一枚十二元的郵票最合宜。如貼兩枚，要平行橫貼，不可上下貼。最多貼三枚。過多則有失美觀。

記在"D"位置上的有很多重要的指示。例如普通寄交主管的信，秘書小姐可以拆後呈閱。但如果有 Private 一字在此處，她就不能拆閱了。

圖十一 信封的四個部份

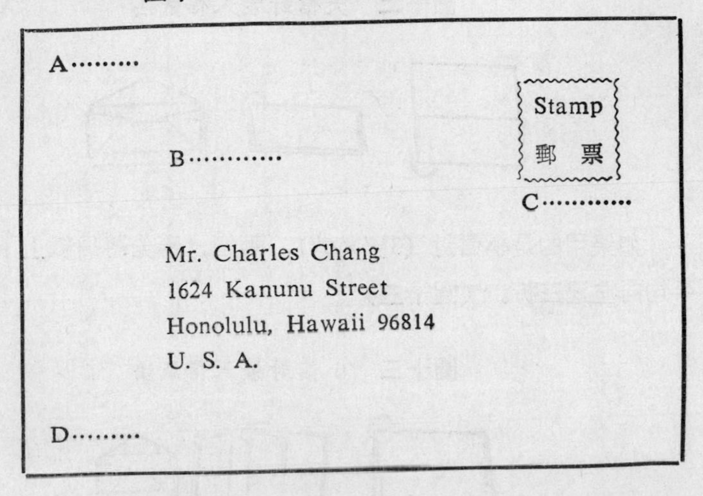

(1) 介紹某先生(Introducing Mr. ......)。

(2) 託某先生帶交(Kindness of Mr. ......)。

(3) 親啓(Private)。

(4) 私函(Personal)。

(5) 密函(Confidential)。

(6) 急件(Urgent)。

(7) 請轉寄(Please forward)。

(8) 留存郵局 (Poste Restante)。

(9) 掛號(REGISTERED)。

第五節 信紙折疊法

凡是用大型信紙($8\frac{1}{2}'' \times 11''$)的，先從下面向內疊起三分之一，再由上面向下疊三分之一，然後裝入大號信封內（圖十二）。也有將信紙的上端，預留一英寸，然後平均疊成三折的。

圖十二 大信封裝入信紙法

如果用的是小信封（$3\frac{1}{2}'' \times 6''$），那麼，應先將信紙上下對疊，再平均向左邊三折。（圖十三）

圖十三 小信封裝入信紙法

另有一種透空式信封，稱窗式信封（Window Envelope）。信封中間有一部份是透明的，從透明部份可以看到信內收信人姓名地址（圖十四）。凡寄發大宗通知書，例如電話收費單等，爲了節省再在信封上打姓名地址，常用窗式信封。

圖十四 窗式信封

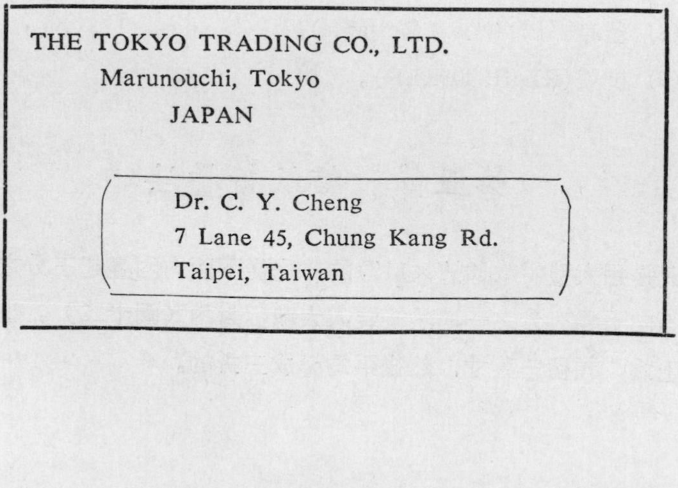

THE TOKYO TRADING CO., LTD.
Marunouchi, Tokyo
JAPAN

Dr. C. Y. Cheng
7 Lane 45, Chung Kang Rd.
Taipei, Taiwan

第六節 商業書信圖解

在敍述了商業書信的九個部份以後，爲了明晰起見，特再將全信圖解如下：

圖十五 商業書信圖解

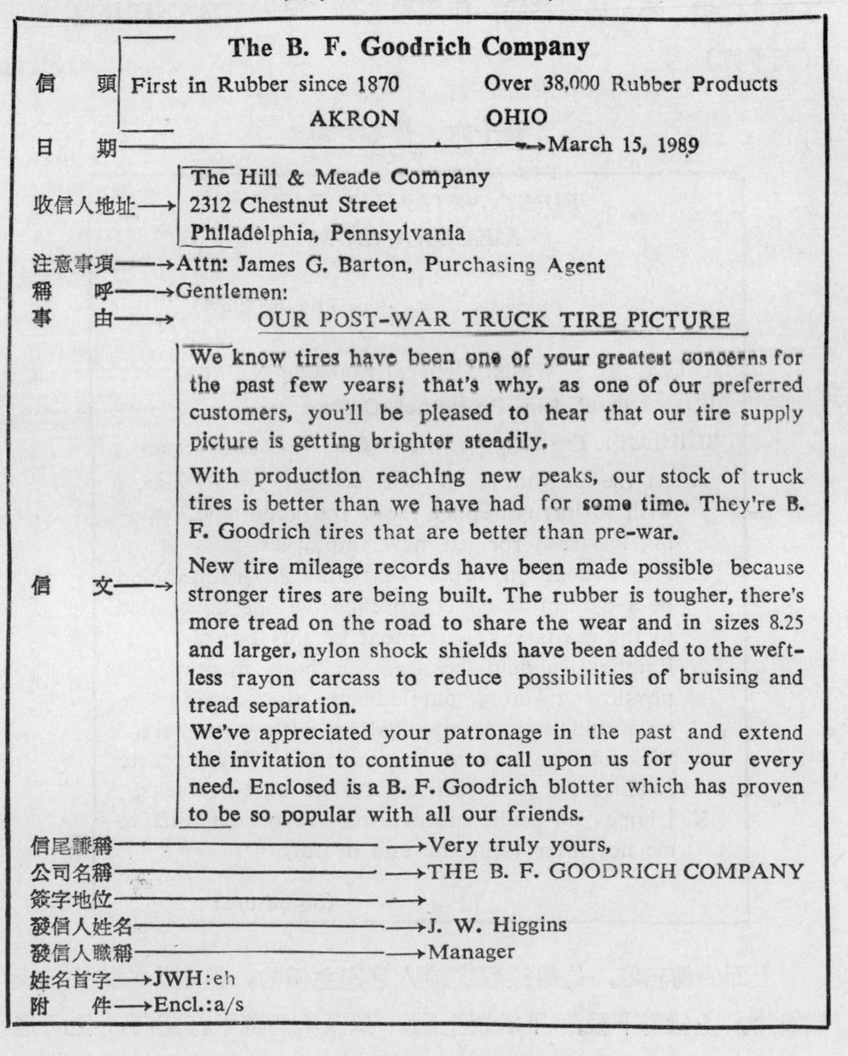

信頭 The B. F. Goodrich Company
First in Rubber since 1870 Over 38,000 Rubber Products
AKRON OHIO

日期 → March 15, 1989

收信人地址 → The Hill & Meade Company
2312 Chestnut Street
Philadelphia, Pennsylvania

注意事項 → Attn: James G. Barton, Purchasing Agent

稱呼 → Gentlemen:

事由 → OUR POST-WAR TRUCK TIRE PICTURE

信文 →
We know tires have been one of your greatest concerns for the past few years; that's why, as one of our preferred customers, you'll be pleased to hear that our tire supply picture is getting brighter steadily.

With production reaching new peaks, our stock of truck tires is better than we have had for some time. They're B. F. Goodrich tires that are better than pre-war.

New tire mileage records have been made possible because stronger tires are being built. The rubber is tougher, there's more tread on the road to share the wear and in sizes 8.25 and larger, nylon shock shields have been added to the weft-less rayon carcass to reduce possibilities of bruising and tread separation.

We've appreciated your patronage in the past and extend the invitation to continue to call upon us for your every need. Enclosed is a B. F. Goodrich blotter which has proven to be so popular with all our friends.

信尾謙稱 → Very truly yours,
公司名稱 → THE B. F. GOODRICH COMPANY
簽字地位 →
發信人姓名 → J. W. Higgins
發信人職稱 → Manager
姓名首字 → JWH:eh
附件 → Encl.:a/s

第七節 備忘錄

除了書信以外，大的公司行號，如銀行之類，在內部通信時，爲了簡化手續，不用書信格式，而用備忘錄（MEMORANDUM）。（圖十六）

圖十六 備忘錄

FIRST NATIONAL BANK
MEMORANDUM

July 6, 1989

FROM: K. S. Wang, General Manager
TO: C. C. Lin, Personnel Officer
SUBJECT: Training Program for New Employees

1. At the beginning of the current fiscal year, I wish to have a report from you about the Training Program for our new employees.
2. Please bear in mind that this Program should be a dynamic and comprehensive one dovetailed to the realistic requirement of this Bank.
3. Emphasis should be laid on both mental and physical training particularly the appropriate ways and manners in dealing with customers.
4. We must get the most competent instructors to conduct the said training for at least 8 weeks.
5. I hope you could submit your report on this to me not later than the end of July.

(Signature)

上面的備忘錄，是總經理寫給人事室主任的。備忘錄的格式，大都如此。上級對下級，下級對上級，以及對內部平行部門，全可適用。例如業務處長有事通知會計處長，人事室主任有事簽報總經理，都可一律用備忘錄。事由下面，再分項說明內容，文字以簡明為主。

習　題

(一) 試用一簡單英文函，說明閉塞式，半閉塞式，和改良閉塞式的各別特點。

(二) 試簡述商業書信的九個部份。

(三) 寫商業書信所使用的文具，應如何選擇?

(四) 試解釋 "P. S.", "a/s", 和"C. C."的意義。

(五) 信封上的 A, B, C, D 部份，作何用途?

(六) 信紙放入大信封及小信封時，應如何分別折疊?

(七) 試述備忘錄的用途，並試擬一簡單的備忘錄。

第四章　商業書信分類實例

第一節　建立貿易關係書信

ESTABLISHMENT OF TRADE RELATIONS LETTERS

提　示

(1) 推展國際貿易，必須"主動"和國外商行多方聯繫，不能守株待兔。

(2) 隨時留意中英文報章雜誌內的"貿易機會"(Trade Opportunity) 專欄。凡與你所經營的行業有關的，都可去信和它們聯繫。

(3) 也可以寫信給外國的商會或貿易刊物，請他們為你推介。

(4) 信內文字要謙遜委婉，說明你所經營業務的性質、歷史、特點等，以及希望建立貿易關係，謀求相互間的利益。

(5) 因為這是和對方往來的第一封信，所以不但內容要文字清通流暢，更要注意信的外觀，如格式、標點、符號、拼法等，所用信紙信封，也要力求精美，讓對方首先留下一個好印象。

(一) 由中國生產力中心推薦

Dear Sirs:

We are much indebted to China Productivity Center for the name and address of your Company and are pleased to learn that you are one of the main producers of the Canned Asparagus in Taiwan.

We are now in the market for this commodity and shall appreciate your quoting us therefor either on FOB Keelung or on C & F Hamburg basis at your earliest convenience. For the moment, we are in need of approximately 20 metric tons each of Tips & Cuts, Center Cuts, and End Cuts to be packed in export cartons of 48 cans each. Since consumers at our end are very critical about quality, you should quote us of Al quality.

As there are many other competing brands scrambling for the consumers' dollars in our market, it is, therefore, imperative that the prices you quoted have to be very competitive.

We look forward to receiving your quotation very soon.

Faithfully yours,

註 ①indebted to 感謝

② China Productivity Center 中國生產力中心

③canned asparagus 罐頭蘆筍　④commodity 貨物

⑤appreciate 感謝　⑥FOB 船上交貨價

⑦C&F 貨價及運費　⑧Hamburg 漢堡(西德大港)

⑨at earliest convenience 便中儘早　⑩approximately 大約

⑪metric ton 公噸　⑫Tips & Cuts 上段

⑬Center Cuts 中段　⑭End Cuts 下段

⑮carton 厚紙盒　⑯consumer 消費者

⑰critical 挑剔 ⑱Al (讀 AONE) quality 最佳品質

⑲brand 牌子 ⑳scramble 爭取

㉑imperative 必須 ㉒competitive 能競爭的

㉓look forward to 期待 ㉔quotation 報價

(二) 由英文中國日報介紹

Gentlemen:

In a recent issue of the "China News", we saw your name listed as being interested in making certain purchases in Taiwan.

We take this opportunity to place our name before you as being a buying, shipping and selling agent. If you do not have anyone here to look after your interests in that capacity, we should be glad if you give us your kind consideration.

We inform you that we have been engaged in this business for the past 25 years. We, therefore, feel that because of our past years' experience, we are well qualified to take care of your interests at this end.

Further, as for references, we can give you from the names of some concerns in your country and also our bankers are Bank of Taiwan, Taipei.

We look forward to receiving a few lines from you in acknowledgment of this letter and with thanks.

Yours, faithfully

註 ①recent issue 最近一期 ②purchase 採購

③opportunity 機會 ④agent 代理人

⑤capacity 地位
⑥consideration 考慮
⑦engage 從事於
⑧experience 經驗
⑨interest 利益
⑩at this end 在這邊
⑪further 再則
⑫reference 諮詢人
⑬concern 有關者
⑭acknowledgment 承認收到
⑮in advance 預先

(三) 由中國郵報推薦

Gentlemen:

The "China Post" has recommended your firm as being interested in establishing business relations with a Chinese firm for the purpose of selling various products of your country and importing Chinese manufactured goods.

We specialize in Central Europe trade, but we have had no contact with your country. We address this letter to you in order to ascertain whether cooperation to the advantage of both our firms could be established.

We invite you to send us details and prices, possibly also samples, of such goods as you would be principally interested in selling, and we shall gladly study the sales possibilities in our market.

On the other hand, please favor us with a list of those articles you are interested in obtaining from here, so that we might be in a position to quote on same and give you all the necessary information regarding supply possibilities.

We enclose herewith a list of goods for which we are representing the leading Chinese manufacturers for their exports or which we are now in a position to supply you from prime sources at the lowest possible prices. This list consists of only a limited selection of the goods which our various departments are regularly supplying, but we are fully equipped to handle your needs for any merchandise whatsoever.

We shall look forward to the pleasure of hearing from you on the above.

Yours faithfully,

註

①recommend 推薦	②firm 公司或行號
③for the purpose of 爲了	④various 各種
⑤manufactured goods 製造品	⑥specialize 專長於
⑦contact 聯絡	⑧address 寫
⑨in order to 爲了	⑩ascertain 確定
⑪cooperation 合作	⑫detail 詳情
⑬sample 貨樣	⑭principally interested主要地有興趣
⑮sales possibility 推銷可能性	⑯on the other hand 另一方面
⑰favor 惠賜	⑱article 貨物
⑲in a position 可以	⑳same 上述事物
㉑information 資料	㉒supply possibility 供給可能性
㉓representing 代表	㉔leading 主要的
㉕prime source 直接來源	㉖consist 包含
㉗limited selection 少數選擇	㉘department 部門
㉙fully equipped 準備齊全	㉚handle 處理
㉛merchandise 貨物	㉜whatsoever 不論何物

(四) 由臺灣銀行國外部介紹

Dear Sirs:

We owe your name to The Foreign Department of Bank of Taiwan, through whom we learned that you are the manufacturers of Textiles, Piece goods and other General Merchandise, and also that you are Importers & Exporters.

May we introduce ourselves as Importers of all General Merchandise, Exporters of Taiwan Produce, and Manufacturers' Representatives, and Commission Agents.

We have been in business since 1923, and can boast of having vast and wide experience in all the lines we handle.

Our bankers are Bank of Taiwan, and The Hongkong & Shanghai Banking Corporation of Hong Kong, from whom you will be able to obtain all the information you may require in regard to our business integrity and financial standing.

We shall thank you to let us know your trade terms, etc., and forward samples and other helpful literature, with a view to getting into business in near future.

Hoping that this letter will be a forerunner to many years of profitable business to both parties, and looking forward to the pleasure of hearing from you.

Yours faithfully,

註 ①owe to 感謝　②textile 紡織品

③piece goods 布疋　④general merchandise 普通商品

⑤produce 產品 ⑥commission agent 佣金代理人

⑦boast 誇口 ⑧vast 大

⑨line 行業

⑩Hongkong & Shanghai Banking Corporation 滙豐銀行

⑪integrity 可靠性 ⑫financial 財務的

⑬standing 地位 ⑭trade term 貿易條件

⑮literature 印刷品 ⑯forerunner 先驅

⑰profitable 獲利的 ⑱both parties 雙方

(五) 由臺灣第一商業銀行推薦

Dear Sirs:

Your good name has been passed on to us by The First Commercial Bank of Taiwan.

Our organization in Taiwan has a number of branches throughout the island. We are in a position to offer you a great service.

You possibly have a Chinese Agent. However, we feel that we are in a very strong position with regard to allocation of licences, and therefore can make a very considerable turnover for you on the basis of our acting as your Agent on a 5% commission. All business would be transacted initially on the basis of Irrevocable Letter of Credit.

Assuming that you have a Chinese Agent, who may or may not be in a strong position, you may not be able to judge whether he is receiving a fair share of the market, unless you have other figures for comparison. Therefore, we suggest that you have your existing Agent, and make this firm one of your Agents. We are quite sure that you will not be disappointed, and if we cannot equal the turnover of your existing Agent, we would be quite happy to cancel our Agency Agreement.

We feel that this is an excellent opportunity for you to strengthen your position here with regard to selling and buying, and we welcome your kind suggestions.

Yours faithfully,

註 ①passed on 轉來 ②agent 代理人
③throughout the island遍及全島 ④with regard to 關於
⑤allocation 分配 ⑥licence 證照
⑦turnover 週轉 ⑧on the basis of 基於
⑨transact 交易 ⑩initially 最初
⑪Irrevocable Letter of Credit 不能收回的信用狀 ⑫assuming 假設
⑬fair share 公平的一份 ⑭figure 數字
⑮comparison 比較 ⑯suggest 建議
⑰existing 現有的 ⑱disappointed 失望
⑲cancel 取消 ⑳Agency Agreement 代理合約
㉑strengthen 加強

(六) 由友人張君推薦

Gentlemen:

Mr. Charles Chang of Gibb Co., Inc., New York, commended your firm to us as one of the largest importers and exporters in Germany.

It so happens that we are not represented in your country at the present time, and we are very much interested in having an active and reputable agent there.

As for ourselves, we are a leading and old established firm of exporters, and we are in a very good position to supply most grades of canned mushroom at competitive prices and for good delivery, and we should be pleased to send you our offers by cable or airmail upon receipt of your specifications covering any grades of mushroom in which you are interested at present.

We also are interested in the import of German papers, and if you are in a position to offer us anything along these lines, we should be pleased to have you send us your quotations, with all pertinent information and with covering samples.

We are looking forward to receiving your reply, and hope we shall be able to establish a connection between our two firms which will be pleasant and mutually beneficial.

Yours very truly,

註 ①commend 稱讚 ②so happen 剛巧
③active 活躍的 ④reputable 有聲譽的
⑤as for 至於 ⑥grade 等級
⑦canned mushroom 罐頭洋菇 ⑧competitive price 能競爭的價格
⑨good delivery 交貨妥速 ⑩pleased 高興
⑪offer by cable 電報報價 ⑫specification 規格
⑬at present 現在 ⑭quotation 報價
⑮pertinent 有關的 ⑯connection 聯繫
⑰pleasant 愉快 ⑱mutually beneficial 互相有利益

(七) 由臺北商會介紹

Dear Sirs:

The name of your esteemed firm has been given us by the Chamber of Commerce of Taipei.

We wish to obtain a supply of fine teacups and saucers, coffee cups and saucers. The type we require is of good quality china and of different shapes, fully decorated with nice flowers or other designs.

We are also interested in good quality cut-glass ware. For your information, the Chinese Government's Import Regulations require that the cutting (that is the cut design) must be equal in value to 10% of the total value of the glassware, and such must be stated in your invoices.

If you can supply this type of merchandise, kindly reply by air mail enclosing your price-list C. I. F. Keelung, Taiwan, and as many illustrations as possible. To facilitate our purchasing, it would be best for you to send us a sample cup by air mail so that we may examine the quality.

In the event of us placing an order, we shall open an irrevocable letter of credit in your favour for the full C. I. F. value of the goods. For your reference, our bankers are Bank of Taiwan, Taipei.

We await your offers with interest.

Yours very truly,

註 ①esteemed firm 貴號　②chamber of commerce 商會
③saucer 茶碟　④china 瓷器

⑤shape 形式 ⑥decorated 裝飾
⑦design 設計 ⑧cut-glass 刻花玻璃
⑨ware 器皿 ⑩regulation 條例
⑪cutting 刻花 ⑫state 聲明
⑬invoice 發票 ⑭price list 價目單
⑮CIF 貨價保險費及運費在內 ⑯illustration 圖畫說明
⑰facilitate 便利 ⑱examine the quality 審查品質
⑲in the event of 倘使 ⑳place order 定貨
㉑in your favor 以你爲受益人

(八)請外國商會代為推薦

Dear Sirs:

We hereby ask your Chamber of Commerce to kindly publish our name and address to the notice of your firms and manufacturers through the medium of your weekly and monthly journals or bulletins, to enable them to communicate with us direct on business transactions.

(a) We wish to point out that we are the importers of the following goods: cottons, woolens and other textile goods, sweaters, cutlery, hats, toys, rayon and silk goods, locks, sewing machines, hosiery, fountain pens, bicycle tyres and inner tubes, bicycles, glassware, earthenware, leather goods, provisions, fancy goods, stationery, jewellery, cameras, and raincoats. Samples, catalogues and quotations are required in each item before we forward our big orders.

(b) We are the exporters of the following articles: —canned pineapple, plywood, and handicraft goods. Samples and quotations will be sent on application.

(c) Terms of payment:—By irrevocable letter of credit.

(d) All quotations should be in U. S. dollar.

We look forward to your kind and favourable reply.

Yours faithfully,

註 ①hereby 藉此 ②publish 公布
③medium 媒介 ④weekly 週刊
⑤monthly journal 月報 ⑥bulletin 公報
⑦communicate 通信 ⑧transaction 交易
⑨sweater 毛線衫 ⑩cutlery 餐具
⑪rayon 人造纖維 ⑫sewing machine 縫級機
⑬hosiery 襪類 ⑭inner tube 內胎
⑮earthenware 陶器 ⑯provisions 食物
⑰fancy goods 新奇美好的貨品 ⑱stationery 文具
⑲jewellery 珠寶 ⑳camera 照相機
㉑raincoat 雨衣 ㉒catalogue 目錄
㉓item 項目 ㉔canned pineapple 罐頭鳳梨
㉕plywood 合板 ㉖handicraft goods 手工藝品
㉗application 申請，索取 ㉘terms of payment 付款條件

(九) 請在會刊內介紹

Dear Sirs:

As we are interested in coming in contact with almost all the leading Importers and General Merchants in your country for the sale of Chinese products at your end, we would request you to kindly insert a note of our desire in your monthly journal under the column of "Opening for Trade", directing them to be put in direct touch with us under your kind auspices.

For your information, we may add that we are one of the largest exporters of handicraft goods, toys, canned asparagus, textile goods, bicycles, etc. If you are to come across with interested members for the purchase of the above products from Taiwan, kindly pass on our name to them, stating that we are in a position to supply their requirements immediately.

We are also interested in coming in contact with exporters and manufacturers of your country for the sale of their products in Taiwan on the exclusive representation. Kindly furnish us with a list of manufacturing concerns and exporters of your country.

We hope our request will have your immediate attention and thank you.

Yours faithfully,

註 ①at your end 在貴地　②request 請求
③insert 插入　④note 啓事
⑤column 專欄　⑥auspices 贊助
⑦come across 遇到　⑧exclusive 獨有的
⑨concern 公司行號　⑩attention 注意

（十）請向當地商號廣為介紹

Gentlemen:

We are writing you these lines to request that you will place our name and address before the interested concerns in your city or country, inserting them in your Bulletin, in regard to purchasing here in Taiwan of a line of rolled, figured and wired sheet glass, both colored and uncolored. This item is used mostly in the large type of buildings, both commercial and industrial. This flat sheet glass is obscure to the vision, but permits light to go through it.

If you will be kind enough to list our name and address as an available source of supply of this building material for importers of this type of glass in your city or country, we would greatly appreciate it.

We should like to state that shipments of this type of glass can be prompt, because we are the sole export agents of one of the largest factories in Taiwan, and have a good quota available for export.

Yours very truly,

註 ①Bulletin 公報 ②rolled 可捲的
③figured 有圖的 ④wired 有鐵絲的
⑤sheet glass 薄玻璃 ⑥obscure 模糊
⑦vision 視線 ⑧permit 容許
⑨available 可有的 ⑩prompt 迅速
⑪sole agent 獨家代理 ⑫quota 配額

(十一) 自我推薦

Dear Sirs:

We introduce ourselves as dealers in bicycles and spare parts. We have been in this line for over two decades. Before the war, we used to import these goods direct from Japan.

We shall be obliged if you will pass on our name to the various manufacturers of these articles and also let us know their addresses.

Thank you.

Yours faithfully,

註 ①introduce 介紹　　②dealer 經銷商
③spare part 零件　　④decade 十年
⑤used to 經常　　⑥obliged 感謝

(十二) 由銀行代客戶探詢貿易機會

Dear Sirs:

At the request of our Bank's valued client, Far East Trading Co., Ltd., Taipei, who are seeking contacts with reputable firms interested in the import of textile goods from Taiwan, we shall be grateful if you will recommend us the names and addresses together with brief credit summaries of a few firms who are willing to contact our client.

On the enclosed sheet, we give you some particulars on our client, which please treat as given in confidence and without any responsibility on our part.

We await your reply with interest and look forward to the pleasure of reciprocating your courtesy whenever the opportunity presents itself.

Yours very truly,

註 ①request 要求 ②valued 有價值的
③client 顧客 ④seeking 尋求
⑤reputable 有聲譽的 ⑥grateful 感激
⑦brief 簡短 ⑧credit 信用
⑨summary 摘要概略 ⑩enclosed sheet 附張
⑪particular 詳情 ⑫treat 作爲
⑬confidence 機密 ⑭responsibility 責任
⑮on our part 在我方 ⑯reciprocate 報答；回報
⑰courtesy 禮遇 ⑱opportunity 時機
⑲present 來到

第二節　詢問貨價和報價書信

INQUIRY AND QUOTATION LETTERS

提　示

(1) 建立貿易關係後，雙方就可互相詢問貨價，準備交易。

(2) 詢價信要詳列所需貨物的名稱、規格、數量等，以便對方答覆，不可籠統含混。

(3) 信中要説明希望交貨的時間，並請早日函覆。

(4) 答覆詢價信，首先應向對方道謝。其次是根據詢問各點一一詳細答覆。價格方面，更要明確，是CIF，或 C&F，或 FOB，或 FAS，必須清楚説明，並告付款方式。

(5) 最後告知所報價格的有效期限，並請早日決定是否同意定購。

(一) 詢問罐頭洋菇價

Re: Canned Mushroom

Gentlemen:

We have recently received many inquiries from retailing shops in the New York area about the captioned item and are sure that there would be very brisk demands therefor on our side. We, therefore, are addressing this letter to you and shall appreciate your quoting us your most competitive prices on a CIF New York basis for the following:

Item	Quantity Required (Case of 24 cans each)
6×68 oz. Stem & Piece	1,500
24×16 oz. Button Slice	1,000
24×16 oz. Whole Slice	2,500
24×8 oz. Button	2,000
48×4 oz. Whole	1,000
48×4 oz. Stem & Piece	1,200

Since this enquiry is an urgent one, please indicate in your quotation the earliest shipment you are able to make for delivery. Please also note that only goods of the best quality are acceptable here. Therefore, you are requested to quote us canned mushroom of the best grade.

We are awaiting your early reply and quotations.

Faithfully yours,

註 ①inquiry 詢問 ②retailing shop 零售店
③captioned 如事由上的 ④item 項目

⑤brisk 興隆的 ⑥therefor 爲此
⑦stem 莖 ⑧piece 片
⑨button 小蕈 ⑩slice 片
⑪indicate 指示 ⑫shipment 貨物交運
⑬delivery 交貨 ⑭note 注意
⑮acceptable 可接受 ⑯grade 等級
⑰quotation 報價 ⑱oz (ounce) 盎斯；英兩

(二) 覆告罐頭洋菇價

Re: Canned Mushroom

Gentlemen:

We thank you for your enquiry about the subject canned food as transmitted by your letter of July 16, 1981 and are very much pleased to know that there are very heavy demands for our products at your end. In the following please find our prices for the six qualities of canned mushroom in which you are interested:

Item	Quantity (Case)	Unit Price CIF net New York per case in U. S. dollar
6×68 oz. Stem & Piece	1,500	12.80
24×16 oz. Button Slice	1,000	13.50
24×16 oz. Whole Slice	2,500	13.40
24× 8 oz. Button	2,000	12.90
48× 4 oz. Whole	1,000	12.80
48× 4 oz. Stem & Piece	1,200	11.20

You would be pleased to know that we have stocks on hand of the first four qualities which can be shipped by the first available vessel sailing to New York direct approximately one week after receipt of your irrevocable and confirmed letter of credit. As for the delivery of the last two qualities, we can deliver them about one month after receipt of your L/C.

You are requested to open the L/C immediately should the prices quoted above be acceptable to you.

Faithfully yours,

註 ①subject 事由所指　②transmit 傳送
③heavy demand 大量需要　④quantity 數量
⑤case 箱　⑥unit price 單價
⑦net 淨價　⑧per 每
⑨stock on hand 現貨　⑩approximately 大約
⑪confirmed 確定了的　⑫L/C 信用狀

（三）詢問糖價

Re: SWC Sugar

Dear Sirs:

We have just received an enquiry from one of our Indian clients who needs 10,000 metric tons of the captioned sugar and shall appreciate your quoting us your best price at an earlier date.

For your information, the quality required should be superior white crystal sugar packed in new gunny bags of 100 kgs. each. Meanwhile, the goods should be surveyed by an independent surveyor as to their quality and weight before shipment. For this enquiry, the buyers will arrange shipping and insurance; therefore, the price to be quoted by you should be on an FAS Kaohsiung basis.

As there is a critical shortage of sugar in India, the goods should be ready for shipment as early as possible. Please be assured that if your price is acceptable, we will place order with you right away.

Your early reply to this enquiry is requested.

Faithfully yours,

註 ①SWC (superior white crystal) Sugar 特級白砂糖

②client 顧客 ③metric ton 公噸

④crystal 結晶體 ⑤gunny bag 蔴袋

⑥kgs. (kilograms) 公斤 ⑦meanwhile 同時

⑧survey 檢驗 ⑨surveyor 公證行

⑩arrange 安排 ⑪FAS 船邊交貨

⑫critical 嚴重的 ⑬shortage 缺少

(四) 覆告糖價

Re: SWC Sugar

Dear Sirs:

We are in receipt of your letter of July 17, 1989 asking us to offer 10,000 metric tons of the subject sugar for shipment to India and appreciate very much your interest in our product.

To comply with your request, we are quoting you as follows:

1. Commodity: Taiwan Superior White Crystal Sugar.
2. Packing: To be packed in new gunny bag of 100 kgs. each.
3. Quantity: Ten Thousand (10,000) metric tons.
4. Price: U. S. dollars one Hundred and Five (US$105.00) per metric ton, FAS Kaohsiung.
5. Payment: 100% by irrevocable and confirmed letter of credit to be opened in our favor through one Al bank in New York and to be drawn at sight.
6. Shipment: Three to four weeks after receipt of letter of credit by the first available boat sailing to Calcutta direct.

Your attention is drawn to the fact that we have not much ready stock on hand. Therefore, it is imperative that, in order to enable us to effect early shipment, your letter of credit should be opened in time if our price meets with your approval.

We are awaiting your immediate reply.

Faithfully yours,

註 ①in receipt of 收到 ②appreciate 感謝
③to comply with 遵照 ④commodity 貨名
⑤packing 包裝 ⑥quantity 數量
⑦price 價格 ⑧payment 付款
⑨shipment 交運 ⑩ready stock 現貨
⑪imperative 必要的 ⑫drawn at sight 見票即付
⑬in time 適時 ⑭approval 同意

(五) 詢混合紡織品價

Dear Sirs:

We are now interested in purchasing substantial quantity of polyester/cotton blended fabrics of the following specifications/constructions and shall appreciate your quoting us your rockbottom prices:

Specification	Blend Ratio	Quantity (yards)
88×75/45's×45's 1/1 57"	65% Polyester/ 35% Cotton	50,000
130×75/45's×45's 1/1 57"	ditto	50,000
110×76/45's×45's 1/1 57"	ditto	50,000

Since competition of these fabrics is very keen hore, it is necessary for you to quote us most competitive prices in order to consummate business.

Your prompt attention to this enquiry will be appreciated.

Faithfully yours,

註 ①substantial quantity 大量 ②polyester 人造纖維
③blend 混合 ④fabrics 紡織品
⑤specification 規格 ⑥rockbottom 最低
⑦ditto 同上 ⑧keen 劇烈
⑨consummate 達成

(六) 覆告混合紡織品價

Subj.: Blended Fabrics

Dear Sirs:

We acknowledge with thanks the receipt of your letter of July 20, 1989 and are pleased to know that you are interested in the captioned fabrics.

To comply with your request, we are submitting our Quotation No. 8-59-1 for your consideration and are pleased to inform you that all the three qualities you enquired about are our specialties and we have sold large quantity of the same cloths to Scandinavian countries where our products have won the buyers' approval just as well at your end. We also wish to advise you that the prices we quoted are our rockbottom ones and it is certain that they can compete favourably with products of other mills.

We shall appreciate it if you will let us have your reply as early as practicable.

Faithfully yours,

註 ①submit 致送　②specialty 專長
③advise 奉告　④mill 工廠

(七) 詢柳安合板價

Re: Lauan Plywood

Dear Sirs:

As a general contractor in Hongkong, we are in need of one million square feet of the captioned goods for prompt delivery. Since you are one of the well-known producers of this in Taiwan, we shall appreciate your quoting us your best price therefor as early as practicable. For your reference, we are giving hereunder the details of this enquiry:

Specification: Lauan Plywood, Rotary Cut, Type III, 3-ply

Size: 1/8″×3′×6′

Quantity: 1,000,000 sq. ft.

Price: Either FAS Kaohsiung or CIF Hong Kong

Shipment: Prompt delivery

Payment: By irrevocable and confirmed letter of credit

It will be greatly appreciated if you will let us have your quotation at the earliest.

Faithfully yours,

註 ①general contractor 營造業包商 ②one million square feet 一百萬方英尺
③well-known 知名的 ④practicable 可以辦到的
⑤reference 參考 ⑥hereunder 下列
⑦rotary 圓形 ⑧ply 股，疊
⑨size 尺寸 ⑩sq. ft. 方英尺

(八) 覆告柳安合板價

Re: Lauan Plywood

Dear Sirs:

We thank you for your letter of July 30, 1989 enquiring about the captioned building material and take pleasure to enclose our Offer No. UE-1109 for your consideration.

Your attention is drawn to the facts that the price we quoted is on an FOB Kaohsiung basis instead of on either FAS Kaohsiung or CIF Hongkong basis and that our offer will be valid until August 31, 1981. As the demands for plywood have been very heavy of late, your early decision and reply are requested.

We are awaiting your prompt and favorable reply.

Faithfully yours,

註 ①building material 建築材料 ②offer 報價單
③for your consideration 請加考慮 ④attention 注意
⑤valid 有效 ⑥of late 最近
⑦decision 決定 ⑧favorable reply 佳音

(九) 詢棉紗價

Re: Cotton and Blended Yarns

Dear Sirs:

We have just received an enquiry from a local weaving mill who is one of our best clients, asking us to supply them the captioned commodities. As we know you are one of the leading spinners in Taiwan, we, therefore, request you to submit quotations for 300,000 lbs. of 40's/1 yarn of 100% cotton on cones and 200,000 lbs. of 45's/1 blended yarn (65% polyester/35% combed cotton) in hanks on a C&FC 2% Hamburg basis. Meanwhile, please also send us by air parcel two sample cones each of these two qualities for our evaluation.

Should the prices you quoted be acceptable and the quality of your sample cones meets with our approval, we will place orders with you forthwith.

Your prompt reply is requested.

Faithfully yours,

註 ①local 當地 ②weaving mill 織布廠
③spinner 紡紗廠 ④yarn 紗
⑤cone 圓錐體 ⑥lbs. 磅
⑦hank 一捲 ⑧C&FC 貨價運費加佣金
⑨evaluation 評審 ⑩forthwith 立即
⑪place order 定貨

（十）覆告棉紗價

Re: Cotton and Blended Yarns

Dear Sirs:

We have received your letter of July 5, 1989 and are pleased to know that you are interested in the captioned yarns and take pleasure to quote you the following prices for your consideration:

Quantity (lbs.)	Specification	Price C&FC2% Hamburg per pound in U. S. cent
300,000	40's/1 (100% Cotton on cones)	42.00

200,000	45's/1 (65% Polyester/ 35% Combed Cotton in hanks)	68.00

As requested, we are sending you, under seperate cover and by air parcel, two sample cones and sample hanks each of the yarns quoted above for your evaluation. It will be appreciated if you will let us know, at an earlier date, that both the quality of our samples and the prices we quoted are acceptable. As for delivery, we can ship the yarns about three months after your orders are confirmed.

Faithfully yours,

註 ①as requcsted 如尊函所囑　②under separate cover 另包
③by air parcel 用航空包裹寄出④delivery 交貨
⑤confirmed 確定

(十一) 詢臺灣手工藝品價

Re: Handicraft of Taiwan Origin

Dear Sirs:

You would be pleased to know that during the past two decades or so, we have been one of the most important importers of European handicrafts in the United States. Now, we have decided to expand the sources of supply to Southeast Asian countries, including Taiwan.

In order to enable you to quote us the right items, we are giving hereunder the articles in which we are particularly interested:

1. Bamboo products—including, but not limited to, basket,

furniture, mats, screens, bamboo wares, etc.

2. Ceramics.
3. Hat and hatbodies--Hat of different kinds, paper hats, Tacha hats, hatbodies.
4. Embroidery and needle work.
5. Lanterns, Palace Lantern in particular.
6. Artificial flowers, including leis.
7. Insect specimens, especially butterflies.
8. Scrolls, reprinted from Chinese paintings.
9. Rattan furniture
10. Musical instruments.

Since this is the first time we plan to deal in Taiwan handicrafts and since we know nothing about the quality of them, it is, therefore, absolutely necessary for you to submit samples of all the items you are to quote us for our inspection. Orders will be placed only after we have approved the samples received.

We are told you are one of the prominent exporters of handicrafts in Taiwan, and are sure that you will supply us goods of the best quality.

Your early response to this enquiry will be appreciated.

Faithfully yours,

註 ①decade or so 十年左右 ②expand 擴大
③source of supply 供應的來源 ④bamboo products 竹製品
⑤basket 袋 ⑥mat 蓆
⑦screen 簾 ⑧ceramics 瓷器
⑨Tacha hat 大甲帽 ⑩embroidery 刺繡品

⑪needle work 針織品
⑫palace lantern 宮燈
⑬lei 花環
⑭insect 昆蟲
⑮specimen 標本
⑯butterfly 蝴蝶
⑰scroll 卷軸
⑱ratten 籐
⑲musical instrument 樂器
⑳absolutely necessary 絕對必要
㉑inspection 檢查
㉒prominent 著名的
㉓response 回覆

(十二) 覆告臺灣手工藝品價

Subj.: Handicrafts

Gentlemen:

Your enquiry of February 10, 1989 requesting us to offer you the captioned commodities has received our immediate attention and we take pleasure to reply as follows.

You probably would be pleased to know that the items in which you are interested under the ten categories are all our specialties. We are, therefore, able to quote you most competitive prices for all of them and assure you that all the goods to be delivered are of Al quality.

Enclosed please find our Quotation No. EXP/60/07 for your reference. To comply with your request and for your evaluation, we are sending samples of all the items quoted under separate cover and by air parcel. Please let us have your reply soon.

Faithfully yours,

註 ①probably 或能　②category 種類
③specialties 特長　④Quotation 報價單
⑤Al quality 最佳品質

(十三) 詢毛線衫價

Subj: Acrylic Sweaters

Dear Sirs:

Many reports have been received from our selling agents in Cape Town, Durban and Johannesburg that there are very heavy demands for the captioned garments in the Union of South Africa.

From Taiwan Export Directory, we are pleased to know that your mill is the most well-known producer of this item in Taiwan. We, therefore, are writing to you for quotations for Acrylic Sweaters for men and ladies of all three sizes: large, medium, and small. As expected, the quantity of our orders to be placed will be very large, and it is anticipated that the prices you are going to quote would be very competitive. Moreover, since the season is coming soon, early deliveries are absolutely necessary.

When quoting, please let us have your prices on both FOBC2% Keelung and CIFC2% Durban bases. Please be assured that should your prices be competitive, we will place our orders with you and open L/C in your favor in time.

Your prompt reply to this inquiry will be appreciated.

Faithfully yours,

註 ①garment 成衣　②Taiwan Export Directory 臺灣出口指南
③Acrylic 人造纖維　④sweater 毛線衫
⑤size 尺寸　⑥medium 中號
⑦anticipate 預料　⑧season 季節
⑨FOBC 船上交貨加佣金　⑩CIFC 貨價保險運費加佣金
⑪be assured 放心

(十四) 覆告毛線衫價

Subj: Acrylic Sweaters

Dear Sirs:

Your letter of February 10 asking us to offer you the subject garments has received our immediate attention. We are pleased to be told that there are very brisk demands for our products in South Africa at present.

In compliance with your request, we are quoting therefore for your decision:

Commodity: Acrylic Sweaters in different color/pattern assortments.

Size: Large (L), Medium (M), Small (S).

Packing: Sweaters are wrapped in polybags and packed in standard export cardboard cartons.

Price: FOBC2% Keelung per dozen in U. S. dollar

L: 12.50

M: 10.50

S: 9.00

CIFC2% Durban per dozen in U. S. dollar

L: 14.00

M: 12.50

S: 11.50

Shipment: 5,000 dozens each size per month; one month after receipt of L/C.

Payment: 100% by irrevocable L/C to be opened with Al bank in London in our favor and drawn at sight.

As our factory is now being run at full capacity to meet very heavy demands abroad, your early decision is absolutely necessary.

Your prompt response is awaited.

Faithfully yours,

註 ①brisk demand 旺盛需要 ②in compliance with 遵照
③pattern 型式 ④assortment 各式各類
⑤wrap 包紮 ⑥polybag 塑膠袋
⑦cardboard 硬紙 ⑧carton 紙盒
⑨per dozen 每打 ⑩L. M. S. 大號、中號、小號
⑪Al bank 最好的銀行 ⑫run 營運
⑬full capacity 最高能量 ⑭response 答復

(十五) 詢耶誕節裝飾品價

Re: X'mas Decorative Objects

Dear Sirs:

As one of the principal dealers in the captioned items in West

Germany, we are sending you this enquiry and shall appreciate your giving it your prompt attention.

For your information, we need electric bulbs of different colors, bells, and small figures of animals and other decorative objects all made of colored aluminum, for decoration of the Christmas trees. The articles we require should be durable and the colors should be bright and attractive. Since what we need should be of Al quality, we would like to pay higher prices for them if the quality is acceptable.

When submitting your quotations, please do not forget to send us samples of the articles quoted. Moreover, if orders are placed by us, please do your utmost to deliver the goods not later than July 31, 1989 in order to enable us to catch up the X'mas Sales of this year.

We will open Irrevocable L/C in your favor right after our confirmation.

Faithfully yours,

註 ①Re 關於 ②X'mas(Christmas)objcct 聖誕節物品
③principal dealer 主要推銷商 ④captioned item 事由上所列之項目
⑤bulb 燈泡 ⑥small figure 小形
⑦decorative 裝飾的 ⑧aluminum 鋁
⑨durable 耐用 ⑩attractive 吸引人的
⑪acceptable 可接受 ⑫submit 遞送
⑬sample 樣品 ⑭catch up 趕上
⑮confirmation 確定

(十六) 覆告耶誕節裝飾品價

Re: X'mas Decorative Objects

Dear Sirs:

We acknowledge with thanks the receipt of your letter of February 11, 1989 enquiring about the captioned objects and are delighted to know that your company is one of the important importers thereof in West Germany.

To comply therewith, we are enclosing our Quotation No. EXP: 60:83 for all the items you required. From the prices quoted, you certainly will note that they are very competitive. However, the competitiveness of our prices is made possible by our mass production and cost control and not at the expense of quality. Therefore, we can assure you that all the goods to be delivered to you are all of A1 quality.

You would be pleased to know that, owing to delicacy of these objects, we have recently devised a new way to pack them in order to ensure that the goods received by our clients will be in excellent condition. With this new device, the percentage of breakage would be greatly reduced.

We thank you again for your interest in our products and look forward to receiving your early and favorable reply.

Faithfully yours,

註 ①thereof 屬於它的　　②comply 遵照辦理
③therewith 隨後　　④mass production 大量生產
⑤cost control 成本控制　　⑥at the expense of 犧牲

⑦assure 保證　⑧owing to 由於
⑨delicacy 易碎　⑩device 方法
⑪client 顧客　⑫excellent 最佳
⑬condition 情況　⑭percentage 百分數
⑮breakage 破損

（十七）詢美棉價

Re: American Raw Cotton

Gentlemen:

As a principal spinner in Free China, we require 1,000 bales of the captioned cotton with the following particulars:

Grade: Strict Middling
Pressley: 90,000
Micronaire: 3.8 NCL

Please quote the price either on an FAS Gulf Ports basis or on a CIF Keelung basis. We are in urgent need of the cotton and should like these 1,000 bales to be delivered at 250 bales each monthly from August through November, 1981.

We are expecting to receive your earliest reply to this enquiry.

Faithfully yours,

註　①raw cotton 原棉　②spinner 紗廠
③bale 包　④middling 中等貨
⑤pressley 拉力（78,000磅以上爲好棉）　⑥micronaire細度
⑦NCL (no control limited) 無上下限　⑧Gulf Ports 墨西哥海灣港口

⑨expecting 盼望

(十八) 覆告美棉價

Re: American Raw Cotton, SM

Gentlemen:

We acknowledge with thanks the receipt of your letter of July 28, 1989 informing us that you are in urgent need of the subject cotton. To meet your requirements, we are offering as follows:

Specification: American Raw Cotton, Type: TASK Grade: Strict Middling, Staple: 1-1/16", Pressley:90,000 Micronaire: 3.8NCL.

Quantity: One Thousand (1,000) bales of approximately 500 pounds each.

Price: U. S. Cents Twenty-four point seventy-five (US¢24.75) per pound FAS Gulf Ports.

Shipment: October, 1989 through January, 1990 at 250 bales each and every month.

Payment: By confirmed and irrevocable letter of credit to be established in our favor through one Al bank in New York one month prior to each monthly shipment and to be drawn at sight.

We regret to inform you that we are unable to comply with your instruction so far as shipments are concerned because the 1981-1982 new crop would not be available until September and trust that you would agree to our proposed shipping schedules. Meanwhile, we are awaiting your early reply in order to catch up with our shipping schedules.

Faithfully yours,

註 ①SM=strict middling 嚴格的中等貨 ②staple 纖維
③FAS 船邊交貨 ④prior to 在前
⑤drawn at sight 見票卽付 ⑥regret 遺憾
⑦comply 遵照 ⑧instruction 指示
⑨so far as 至於 ⑩crop 收成
⑪available 可以有 ⑫shipping schedule 船期
⑬meanwhile 同時

(十九) 詢滑鈎價

Re: Kanai Traveller

Dear Sirs:

We are very much interested in the captioned traveller and wish to use them at our factories. Therefore, you are requested to send us 500 pieces each of the following types and numbers:

Ni-Ni Traveller O Type-Nos. 6/0, 5/0, 4/0
Super Traveller O Type-Nos. 5/0, 4/0, 3/0
Super Traveller OS Type-Nos. 6/0, 5/0, 4/0

Your prompt transmittance of our enquiry to the manufacturer will be appreciated.

Faithfully yours,

註 ①traveller 滑鈎 ②prompt 迅速
③transmittance 轉達

(二十) 覆告滑鈎價

Subj.: Kanai Traveller

Dear Sirs:

Our agents in Taipei have passed your enquiry of July 1, 1989 about the subject traveller on to us for reply and we appreciate very much your interest in our products.

Just for your trial use, we are now sending you, by surface parcel, 500 pieces each of the three different kinds of traveller gratis. It will be highly appreciated if you will let us have your comments on them after trial use.

For your information, we despatched the travellers on July 13, and trust that you would have already received them in time and in good condition.

Faithfully yours,

註 ①agents 代理商　②surface parcel 平郵(非航空)寄遞之包裹
③gratis 免費　④comment 批評
⑤trial use 試用　⑥despatch 寄出
⑦in time 及時

(二十一) 詢機器零件價

Re: Parts for your Sizing Machine Type A-725

Dear Sirs:

We purchased from your company two sets of the subject machine in 1974 and have been operating them ever since and are pleased to inform you that the machines have rendered us very satisfactory service.

Now, almost seven years has elapsed since the installation of the machines. For replacement's purpose, we need at present a considerable quantity of parts/accessories thereof as per the list enclosed herein. It will be appreciated if your will let us have your proforma invoice, in quadruplicate, for them as early as practicable in order to enable us to proceed with all the import formalities. When quoting, please let us have your prices on both FOB Amsterdam and CIF Keelung bases.

The parts and accessories are urgently needed and we, therefore, are expecting to receive your proforma invoice soonest.

Faithfully yours,

註 ①parts 配件 ②sizing 上膠或上漿
③elapse 時間過去 ④installation 安裝
⑤replacement 補充 ⑥considerable 很大的
⑦accessory 零件 ⑧as per 如同
⑨herein 在內 ⑩proforma invoice估價單(申請進口證用)
⑪quadruplicate 四份 ⑫proceed 進行
⑬formality 形式 ⑭urgently 緊急
⑮by return 函到卽覆

(二十二) 覆告機器零件價

Subj.: Parts for Sizing Machine Type A-725

Dear Sirs:

We acknowledge with thanks the receipt of your letter of July 9,

1989 together with enclosure requesting us to quote for parts for replacements of the captioned machine.

As requested, we are submitting our quotation therefor in triplicate and shall appreciate your placing the order with us as early as possible because we have a large number of backlog to be filled during the next few months.

Your early decision is hereby solicited.

Faithfully yours,

註 ①enclosure 附件 ②as requested 謹遵台囑
③triplicate 三份 ④backlog 積壓
⑤solicit 請求

(二十三) 詢松脂價

Dear Sirs:

You would be delighted to know that, since 1960, we have been one of the main importers of raw materials for plastic and petrochemical manufacturers in Taiwan. We are now in need of the following which, we are sure, are the specialties of your corporation:

Shell Molding Resin
Insulation Varnish Resin
Urea Resin Molding Compounds
Paper Finishing Resin

For the preceding four items, please quote us at your earliest convenience your net prices on a CIF Kaohsiung basis per pound. Moreover, to enable us to apply for the import license required, we should like you to airmail your proforma invoice in triplicate by return.

Your early reply to this enquiry will be appreciated.

Faithfully yours,

註 ①main importers 主要進口商 ②raw material 原料
③plastic 塑膠 ④petrochemical 石油化學製品
⑤shell 貝殼 ⑥molding 塑造
⑦insulating 絕緣 ⑧varnish 光亮
⑨urea 尿素 ⑩compounds 混合物
⑪finishing 完工 ⑫resin 松脂
⑬preceding 前面的 ⑭import licence 進口證

(二十四) 覆告松脂價

Dear Sirs:

We acknowledge with thanks the receipt of your letter of February 13, 1989 enquiring about four different kinds of our chemical products and have given it our immediate attention.

As requested, we are submitting our proforma invoice therefor in triplicate for your consideration. Please note that, since you are one of the important importers of petrochemical products in Taiwan, the prices we offered you are the lowest obtainable anywhere. Despite our low prices, the quality of the goods to be shipped to you is always of the best.

We are awaiting your early and favorably reply.

Faithfully yours,

註 ①chemical products 化學產品 ②obtainable 可得到的
③anywhere 任何地方 ④despite 雖然
⑤awaiting your early and favorable reply 靜候早賜佳音

（二十五）詢小麥價

Re: <u>No. 3 Wheat '72 Crop</u>

Dear Sirs:

We have just received an enquiry from a local flour mill asking us to offer 20,000 metric tons of the captioned wheat.

We learned from the Economic Counsellor Office of the Embassy of the Republic of China in Melbourne that your company is the most reliable and well-known exporter of Australian wheat. As such, we shall appreciate it very much if you will quote us your most competitive price on a CIF Keelung basis as early as practicable.

As the flour mill is a regular buyer of wheat, we can assure you of their repeat orders in future should both the price and the quality be acceptable.

We look forward to receiving your early offer.

Faithfully yours,

註 ①local flour mill 本地麵粉廠 ②Economic Counsellor Office 經濟參事處
③embassy 大使館 ④reliable 可靠的
⑤As such 因此 ⑥repeat orders 續來訂單

(二十六) 覆告小麥價

Subj.: No. 3. Wheat '72 Crop

Dear Sirs:

Your letter dated February 13, transmitting your enquiry for 20,000 metric tons of the captioned wheat has been received with thanks.

As stated in your letter, the buyer will be a regular user of wheat in the years ahead. In view of this, we are quoting you our rockbottom price with a view to consummating this business.

Quantity	Commodity	CIF Net Keelung per M/T in Australian Dollar
20,000 metric tons	Australian Wheat No. 3, '72 Crop	252.00

Since the price of wheat has been fluctuating very much of late, the offer we made herein will remain firm up to February 28 only. Meanwhile, we appreciate your efforts to promote the sale of this staple produce in Free China for us.

Faithfully yours,

註 ①transmit 傳遞　②as stated in your letter 如尊函所說
③years ahead 未來許多年　④fluctuate 波動
⑤of late 最近　⑥remain firm 保持不變
⑦efforts 努力　⑧staple 主要的
⑨produce 出產物　⑩consummate 達成

（二十七）詢人造纖維價

Subj: Polyester Staple Fiber

Dear Sirs:

The Board of Directors of our company has resolved to start spinning blended yarns and weaving blended fabrics at our new mill at Taoyuan from April this year and, consequently, we are now in need of the captioned fibers as raw material.

For your information, we are giving hereunder the particulars of this enquiry:

Specification: "Teijin Tetoron" Brand Polyester fiber
T65, Semi-dull, 1.4 denier

Quantity: 300,000 pounds

Packing: Standard export bale packing

Price: FOB Japanese Port

Shipment: 100,000 lbs. each month, April through June, 1989.

Payment: By irrevocable L/C at sight

As the time to start our new operation is imminent, you are requested to submit your quotation at the earliest.

Faithfully yours,

註 ①Board of Directors 董事會 ②resolve 決定
③spinning 紡 ④blended yarn 混合紗
⑤weaving 織 ⑥blended fabrics 混合布
⑦Taoyuan 桃園 ⑧consequently 結果
⑨particular 詳情 ⑩brand 牌子

⑪fiber 纖維	⑫semi-dull 半暗色
⑬denier 尼龍等細度之重量單位	⑭bale 包
⑮FOB 船上交貨	⑯port 港口
⑰operation 使用	⑱imminent 迫切

(二十八) 覆告人造纖維價

Re: "Teijin" Polyester Staple Fiber

Dear Sirs:

Your letter Ref. No. INQ:60:13, of February 11 has been received. We are pleased to know that your company has decided to produce blended yarns/fabrics in future.

In reply to your inquiry, we take pleasure to quote you as follows:

Commodity & Specifications:	"Teijin" Brand Tetoron T65, Semi-dull, 1.4 denier.
Packing:	Standard export carton packing with fibres wrapped in polybags.
Quantity:	Three Hundred Thousand (300,000) pounds.
Price:	U. S. cents Thirty-eight point seventy-five (US¢ 38.75) per pound FOB Japanese ports.
Shipment:	Total quantity to be delivered by three equal monthly shipments, April through June, 1989.
Payment:	100% by irrevocable letter of credit to be opened one month prior to each monthly shipment through the Bank of Tokyo and to be drawn at sight.
Others:	The delivery of goods is subject to the availability of export license.

In view of the fact that our stock on hand has been quite low during the past few months, your early decision is absolutely essential.

Faithfully yours,

註 ①decided 決定 ②standard export carton 標準出口紙盒
③wrap 包紮 ④polybag 塑膠袋
⑤prior to 在前 ⑥Bank of Tokyo 東京銀行
⑦drawn at sight 見票即付 ⑧subject to 限於
⑨availability 可以有 ⑩export license 出口證
⑪in view of the fact 鑒於事實 ⑫stock on hand 手上存貨
⑬absolutely essential 絕對重要

(二十九) 詢玉米價

Re: Corn 1981 Crop

Dear Sirs:

We have been entrusted by Taiwan Supply Bureau (TSB) for the procurement of the captioned commodity and, since this purchase is an urgent one, we have despatched the following telegram to you on February 11:

"PLEASE CABLE QUOTE 10000 METRIC TONS CORN 1981 CROP SUPERIOR GRADE FAS BANGKOK AND CIF KEELUNG NET PROMPT SHIPMENT AWAITING IMMEDIATE REPLY LETTER FOLLOWS"

With regard to price, you are requested to quote on both FAS Bangkok and CIF Keelung bases as TSB might charter a boat to take delivery of the cargo in Bangkok. For your reference, the corn should be packed in new gunny bags of 100 kgs. each. When this transaction is consummated with you, we will ask TSB to open an irrevocable L/C in your favor promptly.

We are anxiously awaiting your firm offer.

Faithfully yours,

註 ①entrusted 被委託　②Taiwan Supply Bureau 臺灣省物資局
③procurement 採購　④despatch 發出
⑤corn 玉米　⑥crop 收成
⑦superior grade 最佳等級　⑧FAS 船邊交貨
⑨Bangkok 曼谷　⑩letter follows 函詳
⑪charter a boat 包船　⑫cargo 貨物
⑬new gunny bag 新蔴袋　⑭transaction 交易
⑮consummate 完成　⑯promptly 迅速
⑰anxiously 焦急地

(三十) 覆告玉米價

Re: Corn '81 Crop

Dear Sirs:

We are in receipt of your letter of February 13, 1989 advising that you have been entrusted by the Taiwan Supply Bureau to purchase 10,000 metric tons of the captioned commodity. We also confirm the receipt of your telegram of February 11 for the said matter.

Much to our regret, we are unable to offer you now because the Thai Government has prohibited all exporters from making firm offers to Taiwan pending the results of negotiation of the Sino-Thai Corn Agreement scheduled to be held in Bangkok sometime in March.

Despite the preceding, we will quote our most competitive price as soon as we are allowed to do so.

Meanwhile, we appreciate very much your sending us the inquiry.

Faithfully yours,

註 ①said matter 上述之事 ②regret 歉憾
③Thai Government 泰國政府 ④prohibit 禁止
⑤pending 懸待 ⑥negotiation 談判
⑦Sino-Thai Corn Agreement 中泰玉米協定
⑧scheduled 定期 ⑨despite the preceding 雖有上述情事

(三十一) 告蘆筍價

Re: Canned Asparagus., Al Grade

Dear Sirs:

We acknowledge with thanks the receipt of your letter of July 3, 1989 enquiring about the captioned commodity and take pleasure to quote you as follows:

Item	Quantity (M/T)	Unit Price (per carton of 48 cans each) FOB Keelung net in U. S. Dollar	C&F Hamburg net in U. S. Dollar
Tips & Cuts	20	12.00	14.00
Center Cuts	20	11.00	13.00
End Cuts	20	9.60	11.60

Remarks: 1. Packing: Standard export cardboard cartons.

2. Validity: Two weeks from the date hereof.

3 Shipment: Total quantity to be shipped during September, 1981.

4. Payment: 100% by irrevccable and confirmed letter of credit to be opened in our favor two weeks before shipment and drawn at sight.

Please note that the prices we quoted above are the rockbottom ones and that our products would compete favourably with other brands in your market because of their superior quality.

Since our stocks have been low for the past few months and the demands therefore are very brisk right now, we urge you to make decision as early as possible.

Faithfully yours,

註 ①unit price 單價 ②per carton of 48 cans each每盒四十八罐
③Tips & Cuts 前段 ④Center Cuts 中段
⑤End Cuts 後段 ⑥cardboard 紙板
⑦validity 有效期 ⑧drawn at sight 見票即付
⑨rockbottom 最低 ⑩favorably 有利
⑪other brands 其他種牌子 ⑫superior quality 優等品質
⑬stocks 存貨 ⑭brisk 旺盛

(三十二) 詢塗漿用品價

Re: "Jet Size" Sizing Stuff

Dear Sirs:

We read with interest your advertisement as inserted in the Septemter 1989 issue of Japan Textile News concerning the captioned sizing stuff.

We want to use the size on a trial basis and shall appreciate your sending us Proforma Invoice, in triplicate, for 1,000 kgs. each of No. 77 for Sheeting and Jeans and Nos. 100-110 for Poplin and Broadcloths at your early convenience. Please note that the prices you quote should be on an FOB Japanese ports basis. Please also send us additional detailed information and also a few pounds of samples of these two kinds for our examination.

We shall appreciate your giving prompt attention to this enquiry.

Faithfully yours,

註 ①advertisement 廣告　②inserted 插登
③issue 期　④sizing stuff 塗漿用品
⑤on a trial basis 作爲試用　⑥proforma invoice 估價單
⑦triplicate 一式三份　⑧sheeting 被單布
⑨jean 斜紋布　⑩poplin 毛葛
⑪broadcloth 廣幅絨布　⑫additional 外加的
⑬detailed 詳細的　⑭information 資料

(三十三) 覆告塗漿用品價

Subj.: "Jet Size" Sizing Stuff

Dear Sirs:

We thank you very much for your enquiry about the subject stuff as transmitted by your letter of July 2, 1981 and are pleased to enclose our Proforma Invoice for your consideration.

We are producers of the stuff for quite a long time and our brands have been very popular among the end-users in Japan as well as in foreigh countries. Under separate cover and by surface parcel, we have despatched to you sample bags of 50 kgs. Each of these two qualities together with brochures explains the merits of our products.

We look forward to receiving your comments on the samples and your early order.

Faithfully yours,

Encl.: a/s

（附件）

Our Ref. No. PI-503 August 15, 198[illegible]

PROFORMA INVOICE

MARKS & NOS.	PACKAGES	QUANTITY (Kg.)	DESCRIPTION	UNIT PRICE	TOTAL AMOUNT
				FOB Osaka net per kg. in U. S. cent	US$
			"Jet Size" Sizing Stuff		
		1,000	No. 77 for Sheeting & Jeans	25.00	250.00
		1,000	No. 100-110 for Poplin & Broadcloths	30.00	300.00
			Total FOB Osaka		550.00
			Ocean Freight		35.00
			Total C&F Keelung		585.00

Remarks: 1. Packing: In polyethylene bags, each containing 50 kgs.
2. Payment: By irrevocable & confirmed letter of credit.
3. Delivery: Approximately one month after receipt of L/C.
4. Validity: This offer is valid until August 31, 1989.

註 ①Jet Size 噴射式 ②brand 牌子
③popular 出名 ④end-user 眞實用戶
⑤under separate cover 另包 ⑥by surface parcel以平郵(非航空)包裹
⑦together with 連同 ⑧brochure 小册
⑨merit 優點 ⑩encl.: a/s 附件如文
⑪marks 標誌(嘜頭) ⑫description 內容
⑬Osaka 神戶 ⑭Keelung 基隆
⑮polyethylene 人造纖維 ⑯approximately 大約
⑰L/C 信用狀 ⑱validity 有效期

(三十四) 詢家庭用具價

Re: Household Products

Dear Sirs:

We were informed by Overseas Chinese Commercial Banking Corporation that you are the producers of the captioned products.

In order to make us familiarized with your products, we shall appreciate your giving us the technical details of them at your early convenience. If available, please also send us literature, brochures, or leaflets dealing with your products in due course.

We will let you know in which item(s) we are interested as soon as we receive the reference materials.

Faithfully yours,

註 ①household products 家庭用具 ②we are informed 我們被告知
③Overseas Chinese Commercial Banking Corporation 華僑商業銀行
④familiarize 熟悉 ⑤technical details 技術細節
⑥available 可有 ⑦literature 說明書
⑧brochure 小册 ⑨leaflet 單張印刷品
⑩in due course 在適當時期 ⑪item 項
⑫reference material 參考資料

（三十五）覆告家庭用具價

Subj.: Household Products

Dear Sirs:

We thank you for your letter of June 29, 1989 wherein you showed your interest in the subject products.

As requested, we are sending you, under separate cover and by air parcel, two copies each of our leaflet, pamphlet, brochure, and catalogue dealing with our products. Since we do not know in what particular items you would be interested, we are unable to submit quotations along with our publications. You, however, are assured that we will quote our best prices as soon as we are told what products of ours you need.

We have been producing these products for almost half a century and they have been sold throughout the world to the fullest satisfaction of our customers. Therefore, we are sure that both the quality and the workmanship of our products would meet with your approval.

Faithfully yours,

註 ①two copies 兩份 ②pamphlet 小册子
③catalogue 目錄 ④dealing with 有關
⑤publication 印刷品 ⑥century 百年
⑦throughout the world 遍及全球 ⑧workmanship 手藝

(三十六) 詢電扇及電鍋價

Dear Sirs:

We are much obliged to China Export Trade Development Council (CETDC) for the name and address of your company and are pleased to know that you are one of the principal manufacturers/exporters of electric fans and rice cookers in Taiwan.

You would be delighted to know that there is a ready market for these electrical appliances to be explored in Indonesia right now and we should like to place large orders with you provided the prices you quote are very competitive. We should also like to serve as your Sole Agents in Indonesia after business relationship between us has been well established. For the time being, we shall appreciate it if you will quote us your rockbottom prices for 1,000 each of electric fans of different sizes and rice cookers for different number of persons at your earliest convenience.

As for payment, we will open irrevocable letter of credit in your favor for the total value right after orders have been confirmed by us.

Faithfully yours,

註 ①obliged 感謝

②China Export Trade Development Council 中華民國國際貿易協會

③electric fan 電扇　④electric rice cooker 電鍋

⑤ready market 現成市場　⑥appliance 用具

⑦explore 開拓　⑧Indonesia 印尼

⑨provided 倘使　⑩competitive 能和他家競爭的

⑪sole agent 獨家代理　⑫establish 建立

⑬for the time being 暫時　⑭at your earliest convenience便中儘速

⑮as for 至於　⑯payment 付款

⑰order 定單　⑱confirmed 確定

(三十七) 覆告電扇及電鍋價

Dear Sirs:

Your letter of February 10 has been received and we are very much pleased to be informed that you are interested in our products and plan to explore the Indonesian market for them. Your efforts in this respect is much appreciated.

In compliance with your request, we are enclosing our Proforma Invoice for all the items about which you inquired. Meanwhile, we should like to assure you that the goods to be delivered to you are all of the best quality.

Please let us have your early reply should the prices quoted and the terms/conditions enumerated in our Proforma Invoice be acceptable.

Faithfully yours,

註 ①pleased 高興　②effort 努力

③in this respect 關於此事 ④appreciate 感謝
⑤ in compliance with 遵照 ⑥request 請求
⑦terms/conditions 條件 ⑧enumerate 列舉

Encl.: a/s

（附件）

PROFORMA INVOICE

February 23, 1989

No. TCEW-60-125

Dear Sirs:

We are pleased to offer you the following items under the terms/conditions given below:

Quantity	Item		FOBC 3% per dozen in U. S. Dollar	CIFC 3% Jakarta per dozen in U. S. Dollar
	Electric Fans			
	Table fan	12″	60.00	72.00
		14″	72.00	84.00
		16″	80.00	100.00
		18″	96.00	100.00
		20″	120.00	132.00
	Floor fan	20″	200.00	225.00
		24″	240.00	265.00

26″	260.00	285.00
Electric Rice Cookers		
for 6 persons	48.00	60.00
for 8 persons	60.00	72.00
for 10 persons	72.00	84.00
for 16 persons	96.00	108.00
for 20 persons	120.00	132.00

Terms and Conditions:

Packing: Standard export wooden case.

Delivery: Two months after receipt of letter of credit.

Payment: 100% by irrevocable and confirmed L/C to be opened in our favor one month prior to shipment and drawn at sight.

Others: All orders subject to our final confirmation.

註 ①FOBC 船上交貨價加佣金 ②CIFC 貨價加保險費加運費加佣金總價
③Jakarta 雅加達(印尼首都) ④wooden case 木箱
⑤receipt 收到 ⑥prior to shipment 裝運以前
⑦drawn at sight 見票即付 ⑧subject to 限於
⑨final confirmation 最後確定

(三十八) 詢購臺灣產品

Gentlemen:

We are interested in obtaining offerings on such commodities as canned crab meat, mandarin oranges, fish livers for vitamin purposes, and split bamboo screens from Taiwan.

This company has been actively engaged in the importing and

exporting business for over 20 years, and our operations extend to all the major ports and active trading areas throughout the world. Our ability to take care of your inquiries is facilitated by departmentalized operations and a staff of experienced department heads and assistants. Our financial standing enables us to control sources of supply of many commodities in which we actively trade. Our terms are irrevocable letter of credit to be established immediately upon our confirmation of an order, and all offerings are made subject to our final confirmation.

An early reply to this letter will be appreciated.

Yours very truly,

註 ①canned crab meat 罐頭蟹肉 ②mandarin orange 柑橘
③fish liver 魚肝 ④vitamin 維他命
⑤split bamboo screen 竹簾 ⑥operations 營業
⑦extend to 伸及 ⑧throughout 遍佈
⑨facilitate 便利 ⑩departmentalized operations分部營業
⑪staff 全體職員 ⑫financial standing 財務地位
⑬enable 使得 ⑭control sources of supply 控制供應的來源
⑮actively 活躍地

(三十九) 詢購各種貨物

Dear Sirs:

We learned from Mr. P. T. Shen of New York that he made representations to you on our behalf in October last year for various lines of merchandise

Up to date, we have not received any communication from you.

and would be interested to know if you are in a position to offer us any of the following commodities:

Canned pineapple

Citronella Oil

Canned Fish

Canned Crab Meat

Silks

Rayons

Cottons

If you are in a position to offer, it will, of course, be necessary for us to arrange import and export permits with our respective authorities. So far as the import permit is concerned, we shall be obliged to submit a firm offer to our authorities as evidence of availability. However, such offers could be made on the understanding that thay are for licence purposes only, and would require subsequent confirmation if and when licences are granted.

We look forward to hearing from you at an early date, and hope that it may be possible to initiate trade with your organization.

Yours faithfully,

註 ①representation 陳述　②on our behalf 代表我們
③up to date 至今日爲止　④communication 音信
⑤in a position 能　⑥canned pineapple 罐頭鳳梨
⑦citronella oil 香茅油　⑧rayon 人造絲
⑨import and export permits 進口及出口證
⑩so far as import permit is concerned 就進口證而言
⑪submit 致送　⑫authorities 當局
⑬evidence 證明　⑭availability 能有

⑮understanding 諒解 ⑯subsequent 隨後的
⑰granted 准發 ⑱initiate 開始

第三節 推銷貨物書信

SALES LETTERS

提示

(1) 推銷新產品，必須在信內列舉產品的各種優點及優待辦法。
(2) 信中應特別說明新產品和市上已有的相同或類似產品不同之處，作一比較，如此方能引起對方之興趣。
(3) 價目單和說明書或樣品可作為附件，一併寄去，促使對方購買。

(一) 推銷吸塵器

Dear Sirs:

We are sure that you would be interested in the new "Tung-Kwang" Vacuum Cleaner which is to be placed on the market soon. Most of the good points of the earlier types have been incorporated into this machine which possesses, besides, several novel features which have been perfected by years of scientific research.

You will find that a special contrivance enables it to run on slightly more than half the current required by machines of equal capability. Further, most of the working parts are readily interchangeable and, in the event of their being damaged, they are thus easy to replace. Such replacement, however, will not be the result of ordinary wear and tear, as only toughened steel is used in the manufacture of moving parts.

The special advantages it offers will make it a quick-selling line, and we are ready also to cooperate with you, by launching a national advertising campaign. Morevoer, we are ready to assist to the extent of half the cost of any local advertising.

Bearing in mind the rapid turnover which is likely to result, you will agree that the 5 per cent commission we are prepared to offer you is extremely generous.

You will find enclosed leaflets and circulars describing this Vacuum Cleaner and we look forward to your agreeing to handle our product as the sole agent in your district.

Yours faithfully,

註 ①vacuum cleaner 吸塵器　②placed on the market 上市
③incorporate 併入　④novel feature 新特點
⑤scientific research 科學研究　⑥contrivance 機械裝置
⑦current 電流　⑧equal capability 同樣能力
⑨further 再則　⑩readily interchangeable 隨時可以互換
⑪in the event of 如果　⑫replace 補充
⑬wear and tear 磨損　⑭toughened steel 加強鋼
⑮moving parts 活動配件　⑯cooperate 合作
⑰launch 發動　⑱campaign 運動
⑲to the extent 限度　⑳local advertising 當地廣告
㉑bearing in mind 記着　㉒rapid turnover 迅速周轉
㉓extremely generous 十分優待　㉔sole agent 獨家代理
㉕district 地區

(二)推銷新化學產品

Dear Sirs:

We think you will be interested in a novelty for which we have secured the sole patent rights. The newly invented product definitely satisfies a long-standing want, and it can save both time and inconvenience.

After months of scientific research, a new substance, Texterite has been evolved. The new material is unaffected by water and most chemicals, and it is light in weight without being delicate. Further, it readily lends itself to household uses, as the enclosed catalogue illustrates.

In order to popularize these products, all the catalogue prices are subject to a special discount of 10 per cent during the month of September only. In all probability this offer will not be repeated for some time, and we accordingly look forward to receiving an early reply from you, when we should be pleased to demonstrate the complete range of these novelties.

Yours faithfully,

註 ①novelty 新奇之物 ②secured 得到
③patent right 專利權 ④definitely 一定的
⑤long-standing 久已存在的 ⑥want 需要
⑦substance 物質 ⑧evolved 產生
⑨unaffected 不受影響 ⑩chemicals 化學品
⑪light in weight 重量輕 ⑫delicate 易破
⑬illustrate 說明 ⑭popularize 推廣

⑮special discount 特別折扣　⑯probability 可能性

⑰repeat 再有　⑱demonstrate 展示

⑲range 範圍

(三) 推銷牛肉精

Dear Sirs:

A sample of our new beef extract, Vitabeef, has been sent to you today by parcel post, which we hope will reach you in perfect condition.

You will find that it possesses many unique features which definitely place it ahead of its many competing brands. It dissolves readily in water, leaving no trace of sediment. As a result, its digestion presents no difficulty, rendering it particularly valuable in cases of convalescence and general debility. Other points are stressed in the leaflet enclosed.

The many inquiries we have received prove that the public are fully aware of the merits of our new product. There will accordingly be no prolonged holding of stocks, with the loss and difficulties such a practice entails.

After paying due regard to the amount of turnover, you will agree that the 15 per cent trade discount which we allowed is decidedly generous, particularly as settlement is to be effected on a monthly basis.

We look forward to adding you to our other retail friends who are reaping substantial profits from the sale of Vitabeef.

Yours faithfully,

註 ①extract 濃縮物 ②parcel post 郵政包裹
③perfect condition 完美狀況 ④possess 有
⑤unique 獨一無二的 ⑥feature 特點
⑦dissolve 溶解 ⑧readily 隨時
⑨trace 痕跡 ⑩sediment 沉澱
⑪digestion 消化 ⑫rendering 使
⑬convalescence 病後復原 ⑭debility 虛弱
⑮stress 強調 ⑯leaflet 傳單
⑰aware of 知道 ⑱prolong 延遲
⑲holding of stock 存貨擱置 ⑳entail 引起
㉑after paying due regard to 在仔細注意後
㉒turnover 週轉 ㉓decidedly 決然的
㉔generous 優惠 ㉕settlement 結帳
㉖reap 收穫 ㉗substantial 很大

（四）推銷擦鞋油

Dear Sirs:

You will shortly receive by parcel post a generous sample of "Black Diamond Polish" which we regard as a distinct improvement on the preparations sold in the market.

From the catalogue enclosed you will observe that this Polish has many advantages, the most important of which we would like to give below.

"Black Diamond" has the advantage of imparting a brilliant finish with a minimum of effort. Further, this brilliance is maintained throughout the day. Besides, shoes treated with this preparation have been found to outlast others and become waterproof.

We are confident that its effectiveness will render this polish a quick selling line, and we believe that the attractive container will conduce materially to this end.

You will find that our prices are surprisingly low and you will appreciate the special discount of 20 per cent to be allowed on all orders placed before the 15th October, 1981.

Yours faithfully,

註 ①shortly 不久 ②black diamond 黑鑽石
③polish 擦鞋油 ④distinct 顯著
⑤preparations 製品 ⑥impart 給予
⑦brilliant 光亮 ⑧minimum 最少
⑨maintain 維持 ⑩outlast 較耐久
⑪waterproof 防水 ⑫confident 相信
⑬effectiveness 效果 ⑭render 使得
⑮attractive 有引誘力的 ⑯container 容器
⑰conduce 助成 ⑱materially 很大
⑲to this end 向此目的 ⑳surprising 驚人的

(五) 推銷臺灣製地毯

Dear Sirs:

Thank you for your enquiry of March 23. We are very glad to enclose our price list of carpets for your reference. A copy of our catalogue has been airmailed to you separately.

Our company has exported big quantities of carpets to your country,

Canada, England, and other countries of the Common Market. You may be sure that your order of our carpets will prove profitable to you.

You will be interested to note that our yearly production is about 20,000 pieces—the largest of all the carpet manufacturers in Taiwan. And, the prices of our carpets are always lower than others.

Recently we have also received enquiries from Japan, South Africa, and Australia. We are convinced that our carpets will be exported to more countries in the near future.

Last month we produced a carpet of a very attractive design for our customer in West Germany. The carpet is not yet in our catalogue. A color picture is enclosed for your reference. We believe you will find the carpet of the new design interesting, and it will surely find a good market in your country. We will send you our best quotation as soon as we hear from you.

Yours truly,

註 ①carpet 地毯　②for your reference 備供參考
③Common Market（歐洲）共同市場
④piece 條　⑤recently 近來
⑥we are convinced 我們相信　⑦design 設計

（六）推銷殺蟲劑

Dear Sirs:

Being anxious to introduce our productions on an enlarged scale in the German market, we hereby wish to approach your esteemed company in the following matter.

We are a manufacturer of chemical-technical productions and specialize in the make of insecticide of a world-famed quality, a list of which is herewith enclosed for your kind perusal.

We inquire whether you would be willing to take up the sale of our products within a certain district of Germany, purchasing the products for your own account and distributing them in your district or on a commission basis.

We are the leading manufacturer of this line in Taiwan and our products are of an excellent quality. Clients who are accustomed to their use never depart from them again.

Delivery terms: We always have a large stock of products and your kind orders can be effected on receipt.

Payment terms: Cash in port of shipment against shipping documents, or an Irrevocable Letter of Credit opened in our favour.

We hold at your disposal samples of those products in which you are interested and samples are required by prospective customers for testing purposes.

Awaiting your reply with much interest.

Yours faithfully,

註 ①anxious 渴望 ②introduce 介紹
③enlarged scale 擴大範圍 ④approach 接觸
⑤esteemed company 貴公司 ⑥chemical-technical 化學技術的
⑦specialize 專長 ⑧insecticide 殺蟲劑
⑨world-famed 世界有名的 ⑩for your perusal 請加詳察
⑪certain district 某一地區
⑫purchase for your own account 自己買下來

⑬distribute 分配　⑭on a commission basis 按佣金辦法

⑮client 顧客　⑯depart 離開

⑰orders can be effected on receipt 收到定單後卽刻辦理

⑱cash in port of shipment against shipping documents 憑交運單據在起運之港口付現金

⑲Irrevocable Letter of Credit 不能取消的信用狀

⑳at your disposal 任憑處置　㉑prospective customer 未來顧客

㉒testing 試驗

(七) 推銷新產綢料

Subj.: Zephyr Silk

Dear Sirs:

We thank you for your inquiry of August 12, 1989 for the subject silk but regret that we cannot supply this material as its manufacture has no longer been continued. It was only when the entire absence of inquiries led us to believe that it had dropped quite out of favour that we decided to take this step.

We are pleased to inform you that you will find our new "G-M-T" brand even more satisfactory.

The new cloth is considerably finer, with a dull lustre that is most attractive; the popularity it enjoys among the leading manufacturers is proof enough of its value.

You will find enclosed a price list and a full range of patterns. It seems to us that a trial order would make you share our confidence.

We look forward to receiving your favourable response.

Yours faithfully,

註 ①entire absence of inquiry 完全無人詢問
②drop quite out of favor 跌至十分不受歡迎的程度
③take this step 採此步驟
④considerably finer 遠較精美
⑤dull lustre 不顯明的光彩
⑥popularity 風行
⑦proof 證據
⑧range 套，範圍
⑨pattern 型式
⑩a trial order 試定一次
⑪share 分得
⑫confidence 信心

（八）覆對新綢料價格及品質完全滿意

Subj.: Zephyr Silk

Dear Sirs:

We are in receipt of your letter of August 22, 1989 enclosing one copy of your price list and a full range of patterns of "G-M-T Silk" for which we thank you.

After perusal of the price list and evaluation of the patterns, we take pleasure to inform you that both the prices and the quality of the cloth are acceptable to us. Please be assured that, whenever opportunity arises, we will place large orders with you for the cloth as a substitute for the captioned brand.

The price list and the patterns are now being kept by us for the purpose to show them to our prospective customers. We are sure that the "G-M-T Silk" will be popular here in the not distant future.

Yours faithfully,

註 ①perusal 細察 ②evaluation 評估
③please be assured 請放心 ④opportunity 機會
⑤arise 來臨 ⑥substitute 代替品
⑦prospective 未來的 ⑧in the not distant future在不久的將來

(九) 推銷臺製皮靴

Dear Sirs:

We are pleased to inform you that yesterday we sent you, by parcel post, our sample of 108-boots for your perusal which were made by two different boot manufacturers in Taiwan. We have offered our quotation on the boots in our letter of July 16. Strictly speaking, we are sure that both the quality and the price of this item are much better than Japanese goods and can entirely cope with the requirements of your market.

However, we are awaiting your comment on this matter. Kindly place a formal order with us promptly; otherwise, our manufacturers will not be able to make optimum arrangement once their production capacity is getting full.

Your kind attention and early reply will be highly appreciated.

Yours truly,

註 ①by parcel post 用郵政包裹 ②boot 皮靴
③strictly speaking 嚴格來說 ④cope with 應付
⑤comment 批評 ⑥formal order 正式定單
⑦otherwise 否則 ⑧optimum 適當的
⑨production capacity 生產能力

(十) 推銷印刷機器

Gentlemen:

Under separate cover, we are sending you catalogs showing the line of our Presses, Cutters, and Duplex Newspaper Presses. It is impossible for us to send you a price list of these items at this time, as they are constantly changed. Delivery of all the equipment is still a good one year away, and our reason for sending you these catalogs is purely for your information, and to do whatever ground work you can until such time as deliveries actually could be made.

We are not in a position to enter into a Dealer's Agreement at this time; but at a later date, we will contact you regarding the possibility of your handling these products in Taiwan.

The writer lived in China for many years prior to the War, and visited Taiwan on several occasions. I am contemplating a trip to the Orient in early October, and in the event that it is possible for me to stop over in Taipei, I will contact you in person. In the meantime, please feel free to write us on any matter pertainting to printing machinery. However, we do not handle the used equipment of any kind.

Very truly yours,

註 ①under separate cover 在另一包裹　②Press 印刷機
③Duplex Newspaper Press 雙用印報機　④at this time 現時
⑤constantly changed 常常改變　⑥equipment 設備
⑦delivery 交貨　⑧purely 完全
⑨ground work 準備工作　⑩actually 眞正
⑪the writer (本函) 寫信人　⑫prior to the War在世界大戰前

⑬on several occasions 因事幾次　⑭contemplating 計劃，考慮

⑮Orient 東方　⑯in the event 倘使

⑰stop over 停留　⑱in person 親自

⑲in the meantime 同時　⑳feel free 隨意

㉑used 已用舊的　㉒Cutter 切紙機

(十一) 推銷做制服用毛料

Gentlemen:

We have this day sent you via airmail, a representative sample of our Navy Blue All Wool Shoddy, which can be used for any type of uniform which you may be making. This is a very nice stock, and we can offer you, subject to prompt cable acceptance, 500,000 pounds on the basis of 39 1/2¢ per pound, net weight, delivered in Taipei; terms should be 100% Irrevocable Letter of Credit, payment in U. S. curency.

We trust that you will be good enough to advise us immediately by cable if this stock is of interest to you. Please understand that the price quoted you is net to us, and the commissions are to be from your end.

Thank you.

Very truly yours,

註 ①representative sample 代表性的樣子

②Navy Blue 海軍藍色　③shoddy 次等貨

④uniform 制服　⑤stock 現貨

⑥subject to prompt cable acceptance 限於迅速電報答覆接受

⑦net weight 淨重　⑧U. S. Currency 美金

⑨advise 通知　⑩commission 佣金

(十二) 推銷玩具

Dear Sirs:

We learn from your ambassador in Taipei that you are looking for new ideas in toys and we feel we may be able to help you.

We have recently bought on very favourable terms the entire stock of plastic toys of the company whose catalogue is enclosed, and are therefore able to offer a wide range of this company's toys at very low prices. Most of the items listed are in stock and we are prepared to offer them to you at the special discount rate of 60% off catalogue prices on orders received by the end of this month for items then in stock.

We think you will also be interested in our own catalogue of mechanical toys and enclose a copy. From the prices listed we would allow you a special discount of 10% over and above our normal trade discount of 30%. We would, however, stress that these special terms, like those for unsold stock of plastic toys, are open only until 30th June. For all orders received after that date prices will be 10% higher than those we are now offering. All prices stated are for delivery c. i. f. New York.

Our settlement terms are 2½% one month from date of invoice.

Both the catalogues we are sending you include a number of novel toys with which we feel sure you would be delighted and we should be very glad to welcome you as one of our customers. We therefore look forward to the pleasure of a first order from you.

Yours faithfully,

註 ①ambassador 大使　②new ideas 新構想
③favorable terms 優惠條件　④entire stock 全部存貨
⑤plastic toy 塑膠玩具　⑥wide range 廣大範圍
⑦60% off 打四折　⑧then 那時
⑨mechanical toys 機器玩具　⑩normal 普通
⑪stress 強調　⑫settlement terms 結帳條件
⑬novel 新奇　⑭delighted 高興

(十三) 推銷雨衣

Dear Sirs:

Thank you for your enquiry of 15th June. We were glad to receive it and to learn of the enquiries you have had for our raincoats. Our "Aquatite" brand is particularly suitable for warm climates and during the past year we have supplied this to dealers in several tropical countries. From many of them we have already had repeat orders. This brand is popular not only because of its light weight, but also because the material used has been specially treated to prevent excessive condensation on the inside surface.

For the quantities you mention we are pleased to quote as follows:

100	"Aquatite" Coats, men's,	medium	@ $3 ea.	$300.00
100	do	small	@ $2.75 ea.	275.00
100	do women's,	medium	@ $2.50 ea.	250.00
100	do	small	@ $2.25 ea.	225.00
				1,050.00
	less 33⅓% trade discount			350.00
	Net price, f.o.b. New York			700.00

Freight (Taipei–New York)	34.50
Insurance	7.25
	US$ 741.75

Terms: 2½% one month from date of invoice

Shipment: Within 3–4 weeks of receiving order

We feel you may be interested in some of our other products and enclose descriptive booklets and a supply of sales literature for use with your customers.

We look forward very much to receiving your order.

Yours faithfully,

註 ①raincoat 雨衣 ②suitable 合宜
③tropical 熱帶的 ④repeat 多次
⑤specially treated 特殊處理 ⑥prevent 防止
⑦excessive 過度 ⑧condensation 凝結
⑨mention 說及 ⑩medium 中號
⑪descriptive 說明的 ⑫booklet 小册
⑬sales literature 推銷用的說明書 ⑭do(ditto) 同上

（十四）推銷臺灣出口產品

Gentlemen:

Mr. C. T. Wang of the Chinese Trading Co., who is a personal friend of ours, has been good enough to turn over to us your letter dated February 25 addressed to them, and we are very happy to learn that you decided to continue your business activities.

From your letter we note that you are about to import various items and we shall be glad to let you have our quotations for the same.

We have our own textile department which can make all offers to you on woolens, worsteds, wool felts, etc. Our machinery department will make our offers to you on machineries, machine tools, spinning and weaving machines. Our chemical department specializes in all kinds of chemicals and will furnish you with all the quotations for these items. We are sorry to inform you that we do not specialize in raw hides and tanning materials, but we are willing to find the right connections for you.

We shall be glad to handle the items you are able to export, but here too, we shall be grateful to you if you will give us full explanations about the merchandise which you are ready just now, when you could furnish us with samples, quotations, delivery terms as well as the commission you would reserve for us.

We have immediately advised our various departments to send our quotations to you for the items in which you are interested and are only waiting for your reaction to furnish you with the samples. We hope and feel certain that there will be a way to work our connections to our mutual satisfaction.

We trust we shall be able to hear from you very soon.

Very truly yours,

註 ①turn over 轉交　②addressed to them 寫給他們的
③activity 活動　④various 各種的
⑤quotation 報價　⑥textile department 紡織品部
⑦woolens 毛織品　⑧worsted 絨線
⑨machine tools 機械工具　⑩spinning and weaving machine 紡織機
⑪raw hide 生牛皮　⑫tanning 製革

⑬connections 聯絡人 ⑭handle 處理
⑮grateful 感激 ⑯explanation 說明
⑰furnish 供給 ⑱reserve 保留
⑲reaction 反應 ⑳mutual satisfaction 彼此滿意
㉑felt 氈

第四節 定貨書信
ORDER LETTERS

提 示

(1) 決定定購貨物後，應在函內詳述所購的貨名、數量、單價、交貨日期、付款方式等。

(2) 如購買的貨物種類很多， 信內不便詳列， 可另附採購單 (Purchase Order)或合約 (Contract)。

(3) 賣方應於覆函時向買方表示承蒙光顧的謝意，並希望能在該批貨物賣出後使買方獲得厚利。

(一) 定購罐頭洋菇

Re: Canned Mushroom

Gentlemen:

We are pleased to receive your prompt reply to our enquiry of August 15 about the captioned canned food and hereby place our order with you as per our Purchase Order enclosed.

As for payment of this order, we will open an irrevocable letter of credit in your favor right away through The First National City Bank, New York, to cover the total CIF value of this order and shall appreciate your arranging to ship the first four items by the first available boat sailing to New York direct right after your receipt of the L/C.

In order to enable us to evaluate the quality of your product, please despatch one sample tin each of the six qualities by air parcel immediately and, meanwhile, confirm your acceptance of this order as soon as possible.

Faithfully yours,

Encls.: a/s

註 ①prompt 迅速 ②canned food 罐頭食物
③right away 立即 ④available 可有的
⑤enable 使能 ⑥evaluate 評量
⑦despatch 寄 ⑧air parcel 航空包裹
⑨meanwhile 同時

(附件)

PURCHASE ORDER

No. GICL/503

Messrs. Taiwan Food Corporation August 12, 1989
25 Po Ai Road
Taipei, Taiwan

We confirm our agreement on purchase of the following goods:

Description: AI Grade Canned Mushroom of the following six qualities:

A. 6×68 oz. Stem & Piece

B. 24×16 oz. Button Slice

C. 42×16 oz. Whole Slice

D. 24×8 oz. Button

E. 48×4 oz. Whole

F. 48×4 oz. Stem & Piece

Quantity: A. 1,500
(Case) B. 1,000

C. 2,500

D. 2,000

E. 1,000

F. 1,200

Packing: By standard export case of 120 cans each.

Unit Price: CIF net New York per case in U. S. dollar.

A. 12.80

B. 13.50

C. 13.40

D. 12.90

E. 12.80

F. 11.20

Payment: 100% by irrevocable letter of credit to be opened immediately through First National City Bank, N. Y. and drawn at sight.

Delivery: For Items A to D: Prompt shipment

For Items E and F: One month after receipt of L/C.

Shipping Marks: On each and every case, the following shipping mark should be stenciled.

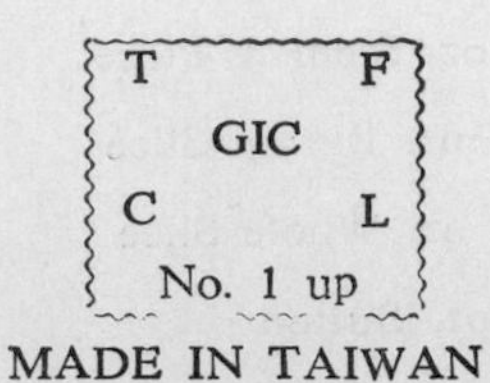

MADE IN TAIWAN

Remarks: 1. Sample cans of each quality to be air-freighted to us for approval.

2. In addition to the ordinary shipping documents, the Seller shall also submit Certificate of Origin for each shipment.

General Importers Co., Ltd.

註 ①Purchase Order 訂購單 ②agreement 協定

③description 說明

④AI grade canned mushroom 最佳等級罐頭洋菇

⑤stem & piece 根及片 ⑥button slice 小蕈片

⑦whole slice 全片

⑧standard export case of 120 cans each 標準出口箱每箱120罐

⑨First National City Bank, N. Y. 紐約花旗銀行

⑩stencile 刷印 ⑪remarks 附註

⑫air-freight 航空貨運 ⑬Certificate of Origin 產地證明書

(二) 允售柳安合板

Re: Lauan Plywood

Dear Sirs:

We thank you for your letter of August 13, 1989 together with one copy of your Order Confirmation No. P/723 for one million square feet of the captioned Plywood and thank you for both.

Enclosed please find two copies each of the original and duplicate of our Contract No. UE-5923 for your signature. Please sign and return one signed copy each thereof for our files.

So far as delivery is concerned, we will try our utmost to ship the total quantity during October 1981, if your L/C arrives here in time. Therefore, in order to enable us to make early shipment, you are requested to arrange the establishment of L/C at your earliest convenience.

We take this opportunity to express our gratitude for your confidence in our company.

Faithfully yours,

Encl.: a/s

註 ①along with 連同　②Order Confirmation 確認定貨單
③plywood 合板，三夾板　④original and duplicate 正本及副本
⑤signature 簽字　⑥signed copy 已簽字的一份
⑦files 卷宗
⑧so far as delivery is concerned 至於有關交貨事宜
⑨try our utmost 試盡我們的最大能力
⑩establishment 開發　⑪gratitude 感激
⑫confidence 信心

(附件)

CONTRACT

Hongkong General Contractors, Ltd.　　Date: August 22, 1989

G. P. O. Box 38

Hong Kong

Contract N. UE-5923

Dear Sirs:

We hereby confirm having sold to you the under-mentioned goods, subject to the terms stated below and conditions specified on the reverse side hereof:

DESCRIPTION: Lauan Plywood, Rotary Cut, Type III, 1/8″×3′×6′/7′, 3-ply, Grading as per JPIC standard.

PACKING: Usual export packing.

QUANTITY: One Million (1,000,000) square feet.

UNIT PRICE: U. S. Dollars Twenty-Nine Point Fifty (US$29.50) per 1,000 square feet, FOB Kaohsiung net.

SHIPMENT: October, 1989

PAYMENT: 100% by irrevocable letter of credit to be opened prior to September 30, 1981 and to be drawn at sight.

Remarks:

Kindly sign and return one copy each of the original and duplicate hereof as evidence of your acceptance.

Confirmed and agreed by: The United Enterprises Co., Ltd.

Hongkong General Contractors, Ltd.

....

(Buyers) (Sellers)

註 ①reverse side 反面 ②Lauan plywood 柳安合板

③as per 如同 ④Kaohsiung 高雄

⑤remarks 附註 ⑥evidence 證明

（三）罐頭洋菇俟信用狀開來即可裝運

Subj.: Canned Mushroom

Gentlemen:

We are pleased to receive your letter of August 12, 1989 enclosing one copy of your Purchase Order No. GICL 503 for six different qualities of the captioned item.

As advised by our letter to you dated August 3, the first four Items are now ready for shipment. We are now contacting shipping companies to book space of the first available boat sailing from Keelung to New York direct. Therefore, you are requested to have the letter of credit opened at the earliest in order to enable us to effect prompt shipment. As for the two other qualities, we can ship them about one month after receipt of your L/C.

As requested, we have already air-freighted one sample tin each of these qualities for your evaluation and, as our products are well received everywhere in the Free World, we are sure that the quality would meet with your approval, too. Meanwhile, we take this opportunity to thank you for the first order you have placed with us and look forward to receiving more orders from you if this first one is executed to your fullest satisfaction.

Faithfully yours,

註 ①shipping companies 輪船公司 ②book space 定位
③first available boat 第一艘可以有的船
④effcct prompt shipment 實行迅速啓運
⑤evaluation 評量　　⑥approval 許可

⑦meanwhile 同時　　⑧take this opportunity 借此機會

⑨execute 執行

(四) 催開信用狀

Subj.: "Jet Size" Sizing Stuff

Dear Sirs:

We acknowledge with thanks the receipt of your letter of August 24, 1989 and are pleased to be favoured with your Order for 1,000 kgs. each of Nos. 77 and 100–110 of the captioned stuff.

In order to meet your urgent requirements, we have already prepared the delivery of the stuff and will book shipping space right after we receive your L/C. Therefore, you are requested to expedite the establishment of credit at the earliest.

Faithfully yours,

註 ①to be favoured 被賜　　②stuff 材料

③in order to 爲了　　④meet urgent requirement 應付緊急需要

⑤expedite 趕辦

(五) 蘆筍備妥九月全部可以交運

Subj.: Canned Asparagus, AI Grade

Dear Sirs:

We thank you for your letter of August 12, 1989 placing an order with us for twenty metric tons each of three qualities of the captioned canned food and take pleasure to inform you that we have already started processing this order. Please be assured that we can effect shipment of the total quantity during September provided your letter of credit is received prior to August 25 as indicated in your letter.

When the goods are ready for shipment, we will cable you as soon as shipping space has been booked. Enclosed please find our Sales Confirmation in duplicate for your signature and we shall appreciate your returning to us one signed copy thereof for our records.

We thank you again for your interest in our product.

Faithfully yours,

註 ①placing an order with us 向我們定貨 ②metric ton 公噸
③start processing this order 開始辦理這批定貨
④please be assured 請放心 ⑤provided 倘使
⑥indicate 指示 ⑦cable you 用電報奉告
⑧Sales Confirmation 售貨確認書 ⑨duplicate 兩份
⑩record 紀錄

（六）決定訂購請速製貨

Subj.: Parts for Sizing Machine Type A-725

Dear Sirs:

Your quotations in triplicate for the subject parts have been receiv- ed. We appreciate very much your giving our inquiry prompt attention.

We have now decided to place an order with you for all the parts as itemized in your quotation and are going to apply for approval to import them to the government agencies concerned. In view of the fact that it will take considerable time to obtain such approval, we hereby request you to start manufacturing them forthwith in order that they could be delivered within three to four months.

Your compliance with our request will be appreciated.

Faithfully yours,

註 ①triplicate 一式三份　　②subject parts 事由上所說之配件

③appreciate 感謝　　④attention 注意

⑤itemize 分項開列

⑥government agencies concerned 有關係的政府部門

⑦considerable 很多　　⑧approval 批准

⑨forthwith 立即　　⑩compliance 照辦

(七) 定購紡績器零件

Re: Spinning Machine Parts

Dear Sirs:

We acknowledge the receipt of four copies of your proforma invoice for the captioned parts as transmitted per your letter dated July 28, 1989.

By means of this letter, we confirm our placing this order with you. You would be pleased to know that we have already applied to the government agencies concerned for import license and letter of credit required for this order. We will let you know the reference numbers of both the import license and letter of credit as soon as they are available.

We also thank you for granting us 2.15% discount on your catalogue prices for all the articles you quoted.

Faithfully yours,

註 ①proforma invoice 估價單 ②transmit 傳遞
③per 由 ④by means of 用
⑤have already applied 已申請 ⑥import license 進口證
⑦reference number 參考號碼 ⑧grant 給予
⑨catalogue 目錄 ⑩available 可以有

(八) 定購罐頭鳳梨

Re: Canned Pineapple

Dear Sirs:

We have received your letter of August 15, 1989 along with your Proforma Invoice No. PI-503 for the subject food.

We hereby place an order with you for 1,000 cases each of Nos. 77 and 100 at US$5.00 and US$6.00 per case FOB Keelung respectively. For your information, we have already applied for import license and letter of credit for this order and will inform you the reference numbers of both as soon as our applications have been approved by the government agencies concerned.

Since we need the goods quite urgently, you are requested to effect shipment one month after receipt of our L/C.

Faithfully yours,

註 ①along with 連同 ②per case 每箱
③respectively 各別地 ④for your information 茲特奉告
⑤quite urgently 十分緊急 ⑥effect shipment 交運

（九）購買柳安合板

Subj.: Lauan Plywood

Dear Sirs:

We are pleased to receive your letter of August 3, 1989 together with one copy of your offer No. TH-1109 for one million square feet of the subject commodity.

Upon perusal of the terms of your offer, we have found them acceptable and enclose our Purchase Order No. PO/723 in duplicate for your signature. Please sign and return one signed copy thereof for our files.

As we are in urgent need of the building materials, you are requested to effect shipment during October, 1981 as promised in your offer. Meanwhile, we are to apply for the establishment of an inrevocable letter of credit with your Company as the beneficiary within the next few days.

Your prompt confirmation is awaited.

Faithfully yours,

Encl.: a/s

註 ①million 百萬　②commodity 貨物
③perusal 細審　④acceptable 可接受
⑤purchase order 購貨定單　⑥duplicate 一式二份
⑦signature 簽字　⑧signed copy 已簽字的一份
⑨beneficiary 受益人

（附件）

Order Confirmation

Order No. PO/723

The United Enterprise Co., Ltd.

P. O. Box 8870

Taipei

Dear Sirs:

We confirm having purchased from you the following commodity:

1. Commodity &
   Specification: Lauan Plywood, Rotary Cut, Type III,
   1/8", 3-Ply, Grading as per JPIC standard.
2. Size & Grade: 1/8"×3'×6'/7'
3. Quantity: One Million (1,000,000) square feet.
4. Unit Price: U. S. Dollars Twenty-nine Point Fifty (US$29.50) per 1,000 square feet, FOB Kaohsiung.
5. Shipment: Total quantity to be shipped during October, 1989
6. Payment: 100% by irrevocable letter or credit to be opened in early September and to be drawn at sight.
7. Survey or Inspection: Inspection at the Seller's mill to be final.

Hongkong General Contractors, Ltd.

註 ① specification 規格　②Lauan plywood 柳安合板
③rotary 圓形　④ply 股，叠，層
⑤size 尺寸　⑥grade 等級
⑦unit price 單價　⑧survey 檢驗
⑨inspection 檢查　⑩mill 工廠

(十) 定購罐頭蘆筍

Subj.: Canned Asparagus, AI Grade

Dear Sirs:

We are in receipt of your letter of August 3, 1989 submitting your quotations for three different qualities of the subject commodity.

In view of the fact that the prices you quoted are acceptable, we hereby place the following order with you:

Quality	Quantity (M/T)
Tips & Cuts	20
Center Cuts	20
End Cuts	20

So far as prices are concerned, we prefer C&F Hamburg to FOB Keelung. It is understood the prices for the former are US¢75.00, 71.00 and 44.00 per kg. for these three qualities respectively.

As requested in your letter, we will establish an irrevocable letter of credit in your favor around August 20, in order to enable you to effect shipment during September, 1989.

Please let us have your confirmation at the earliest.

Faithfully yours,

註 ①canned asparagus 罐頭蘆筍 ②Al grade 最佳等級
③submit 送 ④commodity 貨物
⑤M/T 公噸(metric ton)
⑥so far as prices are concerned 至於有關價格方面
⑦prefer C&F Hamburg to FOB Keelung 比較願意貨價加運費在漢堡

交貨而非在基隆船上交貨（即不包括運費的貨價）

⑧former 前者　　⑨US¢ 75 美金七角五分

（十一）購臺灣砂糖一萬頓

Subj.: SWC Sugar

Dear Sirs:

We acknowledge the receipt of your letter of August 1, 1989 offering us 10,000 metric tons of the captioned sugar and wish to confirm this order with the following particulars:

Commodity: Taiwan Superior White Crystal Sugar.

Packing: To be packed in new gunny bag of 100 kgs. each.

Quantity: Ten Thousand (10,000) metric tons.

Price: U. S. dollars One Hundred and Five (US$ 105.00) per M/T FAS Kaohsiung net.

For your information, we have already applied for the opening of an irrevocable letter of credit in your favor to Chase Manhattan Bank, New York and trust you would receive it within one week. As stated in our letter to you dated July 17th, it will be highly appreciated if you will effect shipment of these 10,000 M/ T's by the first available vessel sailing to Calcutta direct as soon as you receive our L/C. Moreover, since the price contracted is on FAS basis, we shall appreciate your arranging with shipping companies for delivery of the goods with ocean freight to collect.

Faithfully yours,

註　①SWC sugar 特級白砂糖　　②particulars 詳情

③gunny bag 蔴袋 ④FAS Kaohsiung net高雄船邊交貨淨價

⑤Chase Manhattan Bank 大通銀行

⑥first available vessel 第一艘可以有的船

⑦Calcutta 加爾各答(印度海港) ⑧arrange 安排

⑨shipping company 輪船公司 ⑩freight to collect 運費到付

(十二) 定購耶誕節裝飾品

Subj.: X'mas Decorative Object

Dear Sirs:

Attention: Mr. C. Y. Cheng, Manager

This serves to acknowledge the receipt of your letter of February 13 enclosing your Quotation No. EXP:60:83, in triplicate, for the captioned objects.

As stated in your letter, we are pleased to have found that the prices you quoted are quite competitive. Besides, we are also pleased to know that you have devised new ways to pack the objects with a view to avoiding damage in transit. In view of these, we are enclosing our Purchase Order No. IMP:71:532, in duplicate for your signature. Please sign and return to us one copy thereof as your confirmation of this order.

Apropos of delivery, it will be highly appreciated if your could arrange to ship the goods in early July in order to enable us to catch up with the Yuletide sales. We will open an irrevocable L/C to cover the total C&F value of this order in due course.

Faithfully yours,

註 ①X'mas 即 Christmas ②decorative object 裝飾品
③serve 作爲 ④triplicate 一式三份
⑤devise 發明，創造 ⑥new ways 新方法
⑦pack 包裝 ⑧in transit 中途
⑨in view of these 有鑒於此 ⑩apropos of 關於
⑪catch up 趕及 ⑫cover 包括
⑬in due course 到相當時期 ⑭Yuletide 耶誕季節

（十三）定購毛線衫

Re: Acrylic Sweaters

Dear Sirs:

Your letter of February 20, 1989 quoting the subject garments has been received and we have given it our immediate attention.

By means of this letter, we place our order with you for the sweaters of the following particulars:

Acrylic sweaters of different colors/patterns assortments:

Large	2,000 dozens
Medium	3,500 dozens
Small	1,500 dozens

As the season is coming soon, you are requested to ship the total quantity of this order not later than June, by which time we will open our irrevocable L/C for the total value thereunder.

Please confirm the shipping schedule as requested soonest.

Faithfully yours,

註 ①acrylic 人造纖維 ②sweater 毛線衫
③garment 成衣 ④by means of this letter 用本函
⑤pattern 型式 ⑥assortment 各式俱備
⑦medium 中號 ⑧irrevocable L/C 不能收回的信用狀
⑨shipping schedule 船期

(十四) 定購電扇及電鍋

Subj.: Electrical Appliances

Dear Sirs:

We acknowledge the receipt of your letter dated February 23rd, along with your Proforma Invoice in triplicate and take pleasure to place our order with you for the following:

Item	Quantity (doz.)
Electric fan	
Table fan 14″	50
ditto 20″	60
Floor fan 20″	80
ditto 26″	70
Electric Rice Cooker	
for 8 persons	40
for 16 persons	60
for 20 persons	80

Since this is the first time we explore the Indonesian markets for your products, it is, therefore, absolutely essential for you to see to it that all the goods delivered are up to the export standards and that all of them are thoroughly inspected prior to shipment.

We will open an irrevocable letter of credit with your company as beneficiary as soon as we receive your advice that the goods are ready for shipment.

Faithfully yours,

註 ①electrical appliances 電器用具
②Proforma Invoice 估價單（多作申請進口證用）
③triplicate 一式三份 ④place order with you 向你定貨
⑤doz＝dozen（打） ⑥table fan 桌扇
⑦floor fan 立扇 ⑧ditto 同上
⑨electric rice cooker 電鍋 ⑩explore 發掘
⑪absolutely essential 絕對重要
⑫up to the export standard 符合出口標準
⑬thoroughly inspected 徹底檢查
⑭prior to 在前 ⑮beneficiary 受益人

（十五）購買手工藝品

Subj.: Handicraft Goods

Dear Sirs:

We thank you for your letter of February 25, 1989 enclosing Quotation No. EXP/60/70 and are pleased to inform you that the samples you sent us have already been received.

In view of the fact that both the prices you quoted and the quality of the samples are all right, we are enclosing our Purchase Order No.

72/71 for your processing. As requested, we will open an irrevocable letter of credit in your favor in time to cover the total value of this order.

As for delivery, we wish to ask you again that the goods ordered have to arrive in New York in early September in order to enable us to catch up the X'mas sales.

Faithfully yours,

註 ①handicraft goods 手工藝品 ②quotation 報價單
③in view of 鑒於 ④all right 很好
⑤processing 辦理 ⑥in time 及時
⑦as for delivery 至於交貨 ⑧in order to 爲了
⑨enable 使 ⑩X'mas 耶誕節

(十六) 定購棉紗

Re: Cotton and Blended Yarns

Dear Sirs:

Your letter of August 3, 1989 quoting the subject yarns has been received and the prices quoted therein are found acceptable.

By means of this letter, we place an order with you for 300,000 pounds of 40's/1 100% cotton on cones and 200,000 pounds of 45's/1 blended yarn (59% Polyester/35% Combed Cotton) in hanks at the respective prices of US¢ 42.00 and US¢68.00 per pound C&F2% Hamburg.

As stated in your letter under reply, the yarns are to be delivered about three months after confirmation of order. Therefore, we expect the shipment of the first batch of yarns to be effected in early November, 1981.

As for payment, we will open L/C to cover the first shipment in early October, 1989.

Faithfully yours,

註 ①blended yarn 混合紗 ②acceptable 可以接受
③by means of 用 ④cone 圓錐體
⑤polyester 人造纖維 ⑥combed cotton 梳過的棉花
⑦hank 捲 ⑧first batch 第一批

(十七) 要求開罐器延期交貨

Dear Sirs:

We thank you for your order of August 5, 1989 for 5,000 Tin Openers but regret that we are unable to execute the order from stock owing to the heavy demand recently for these gadgets.

Our suppliers have, however, undertaken to replenish our stock within seven days and we trust that it will not be inconvenient for you to allow us this extention.

Should you be willing to meet our wishes, we should be grateful if you would confirm your order on the revised conditions.

Yours faithfully,

註 ①tin opeuer 開罐器 ②regret 遺憾
③execute 執行 ④stock 存貨

⑤gadget 設計精巧的小機械　⑥replenish 補充
⑦trust 相信　⑧inconvenient 不方便
⑨extension 延期　⑩grateful 感激
⑪revised 修改了的　⑫condition 條件

(十八) 因罷工請准延期交貨

Dear Sirs:

It is with great regret that we find ourselves compelled to ask for an extension of time in the execution of your order No. 34 of August 7, 1989.

The strike of the workers at our mill has completely disrupted our business and though we have made strenuous efforts to fulfil our obligations, the general nature of the trouble has rendered our attempts in vain

Negotiations for settlement are, however, proceeding apace and we have every reason to believe that we could effect delivery within a fortnight. We trust that this course will meet with your approval and suffice to save you further inconvenience.

We tender our apologies and hope that you will not allow these circumstances over which we have no control to influence you in your judgment of our handling of orders.

Yours faithfully,

註 ①compelled 被迫　②extension of time 展長時間
③execution 執行　④strike 罷工
⑤mill 工廠　⑥disrupt 中斷

⑦strenuous 艱辛的　⑧fulfil 達成
⑨general nature 一般性質　⑩trouble 麻煩
⑪render 使　⑫attempt 嘗試
⑬in vain 無效　⑭negotiation 談判
⑮settlement 解決　⑯proceeding apace 快速進行
⑰fortnight 兩星期　⑱course 方法，途徑
⑲suffice 足以　⑳tender 提出
㉑apology 道歉　㉒circumstance 情況
㉓influence you in your judgment of our handling of orders 影響你對於我們辦理貴處定貨的判斷

（十九）覆允延期交貨

Re: Our Order No. 34

Dear Sirs:

We acknowledge the receipt of your letter of August 16, 1989 concerning the delivery under the captioned order.

It is a regrettable thing that the workers of your mill have gone on strike, thus disrupting completely your production and making you unable to fulfil your obligations to cutomers. Since strike is something beyond your control, we concur in extending the delivery under the captioned order for another three weeks by which time we are sure reconciliation would have been reached between your company and the workers.

As strike is a force majeure, please be assured that we will not judge your business methods unfavorably by such an unforeseen incident and we have confidence in your mill just as heretofore.

Faithfully yours,

註 ①beyond your control 非你所能控制 ②concur 同意
③reconciliation 調解 ④reach 達成
⑤force majeure 不可抗力 ⑥unfavorably 不利地
⑦unforeseen incident 不能預見的偶發事件
⑧confidence 信心 ⑨heretofore 以往

(二十) 補充紡紗機器零件

Re: Spinning Machine Parts

Dear Sirs:

You would remember that, in 1982, we bought from you 50 sets of the captioned machine which are now being operated perfectly well at our mill.

After having run these machines for almost ten years, we need the necessary parts/accessories for replacements. Enclosed is a detailed list itemizing the parts and accessories we need together with their serial numbers and quantities for your ready reference. We shall, therefore, appreciate your airmailing us your Proforma Invoice therefor soonest in order that we can apply for import license and the necessary foreign exchange to pay you.

Please give this enquiry your immediate attention.

Faithfully yours,

註 ①remember 記得 ②operate 使用

③perfectly well 完全良好	④run 使用
⑤parts 配件	⑥accessories 零件
⑦replacement 補充	⑧detailed list 詳單
⑨serial number 連續號碼	⑩ready reference 現成參考
⑪by return 接信後卽覆	⑫import license 進口證
⑬enquiry 詢問	⑭foreign exchange 外滙

(二十一) 覆告紡紗機器零件須四月後方能交貨

Re: Spinning Machine Parts

Dear Sirs:

We are in receipt of your letter of July 10, 1989 asking us to quote for the captioned parts and are pleased to hear from one of our old customers again.

Enclosed please find our Proforma Invoice, in quadruplicate, for the parts you require. Please note that we have given you, our longtime client, a 15% discount on all the items quoted. As for delivery, we regret to inform you that, owing to the fact that our factory is now being operated at 100% capacity, we can only effect shipment four months after the order is confirmed. As this is something beyond our control, we can do nothing but request your understanding.

We are awaiting your early confirmation of this order.

註

①spinning machine 紡紗機	②parts 配件
③old customer 老顧客	④quadruplicate 一式四份
⑤longtime client 爲時很久的主顧	⑥item 項目
⑦as for delivery 至於交貨	⑧owing to 由於

⑨capacity 能力 ⑩understanding 諒解

(二十二) 定購計算機

Dear Sirs:

We thank you for your letter of November 4, 1989.

We have studied your catalog and have chosen 3 models of calculating machine for which we enclose our order. We would stress that this is a trial order and if we are satisfied with your shipment you can expect regular repeat orders.

To avoid difficulties with the custom authorities here, please make sure that our shipping instructions are carefully observed.

For our credit status we refer you to the Transatlantic Bank, Old Bond Street, London and Messrs. Lafayette & Cie, Geneva.

Yours truly,

註 ①model 型式 ②calculating machine 計算機
③stress 強調 ④trial order 試定
⑤regular repeat orders 經常不斷的定貨
⑥avoid 避免 ⑦custom 海關
⑧authorities 當局 ⑨shipping instructions 裝運指示
⑩credit status 信用狀況 ⑪Geneva 日內瓦
⑫observe 遵守

(二十三) 因缺原料歉難如期交貨

Dear Sirs:

We acknowledge the receipt of your letter of November 3 and are extremely sorry that we have not been able to deliver your order in time.

The delay was caused by the belated arrival of some of the raw material. We are glad, however, that the dresses will be ready for shipment next week and we hope that they will arrive in time for the season.

Please accept our apologies for the delay and the inconvenience it has caused you. It was due to reasons entirely beyond our control.

Yours truly,

註 ①extremely sorry 十分抱歉 ②in time 及時
③delay 延誤 ④belated 誤期的
⑤raw material 原料 ⑥dress 衣服
⑦season 季節 ⑧apologies 道歉
⑨due to 由於
⑩entirely beyond our control 完全出於我們的控制以外

(二十四) 因火災無法交貨

Dear Sirs:

We have received your letter of November 5 and are very sorry that we have been unable to supply your order in time. Unfortunately, production in one of our factories was held up for three weeks by a fire and in spite of all our efforts it was not possible to make good the delay.

In the circumstances we have regretfully cancelled your order but trust that you will give us the chance of quoting again when you are in the market next time.

Yours truly,

註 ①unfortunately 不幸地 ②held up 停頓
③fire 火災 ④in spite of 雖然
⑤efforts 努力 ⑥make good 補救
⑦circumstance 情況 ⑧cancel 取消
⑨chance 機會

(二十五) 衣已製成卽可交運

Dear Sirs:

We are glad to say that the dresses which you ordered in July are ready for shipment. We have made special effort to complete your order in time and we are confident that the superb tailoring of these dresses will give you full satisfaction.

Please let us have your instructions for packing and shipment.

Yours truly,

註 ①ready for shipment 準備交運 ②confident 相信
③superb tailoring 上等縫工 ④full satisfaction 完全滿意
⑤instructions for packing and shipment 包裝及交運的指示

(二十六) 希望首次交易雙方滿意

Dear Sirs:

We want to tell you how pleased we were with your order because it represents our first dealing with you. We have always felt that our high quality merchandise should have a ready sale in a fashionable shop like yours.

It is our hope that this first transaction will be the beginning of a long and happy association. You can be sure that we shall do our best to satisfy you.」

Yours truly,

註 ①pleased 高興　②represent 代表
③first dealing 初次交易　④ready sale 良好銷路
⑤fashionable 時髦　⑥transaction 交易
⑦association 往來

(二十七) 要求定貨加價

Dear Sirs:

We acknowledge with thanks your order No. 1109 for delivery June/July.

We regret, however, that we cannot book the order at the prices we quoted 6 weeks ago. As you know, wages and materials have risen substantially in the meantime and we were reluctantly compelled to adjust our prices in order to cover at least part of this increase.

The lowest prices we can quote today are as follows:

Model A	$ 14. 50
Model B	$ 18. 30
Model C	$ 35. 25

We do not want to influence you, but we think it only fair to mention that we shall have to increase these prices substantially again when our old stock of material is used up.

Please inform us soonest whether we may book your order at these prices. We shall then be able to give you delivery in June/July as required.

Yours truly,

註 ①book the order 登記定單 ②wage 工資
③material 原料 ④have risen 已漲
⑤substantially 大量地 ⑥in the meantime 同時
⑦reluctantly 不得已 ⑧compelled 被迫
⑨adjust 調整 ⑩cover 包括
⑪at least part of the increase 至少增加的一部份
⑫ influence 影響 ⑬fair 公平
⑭mention 說 ⑮have to 必須
⑯used up 用完

(二十八) 同意加價

Dear Sirs:

In reply to your cable of June 6 and your letter of the same date, we have pleasure in enclosing our order No. XY/3312 for your prompt and careful attention.

We have accepted your price of $14.50 for Model A and $18.30 for Model B, but must ask you to keep your quotation for Model C to $30.25. Quite frankly, we think that even at this price it will be difficult to sell as there are so many cheaper models on the market.

Payment will be made by Letter of Credit in New York against documents.

Please advise us when the goods are ready for shipment and await our final shipping instructions.

Yours truly,

註 ①letter of the same date 同日的信 ②prompt 迅速
③quite frankly 十分坦白地說 ④cheaper models 較廉的型式
⑤payment 付款 ⑥against document 憑單據

(二十九) 定購雜貨貨到付款

Gentlemen:

Please fill the order for the sundry goods as mentioned in the enclosed order sheet at your early convenience. You may send the goods to me C. O. D.

We have not seen a copy of your latest catalog, but we trust that the goods we ordered can be identified by the numbers given therein. Especially please pay your particular attention to the packing so that the goods may not be damaged in conveyance.

Very truly yours

註 ①fill the order 辦理定貨 ②sundry goods 雜貨
③order sheet 定單 ④C. O. D. 貨到付款
⑤latest 最近的 ⑥catalog 目錄
⑦identify 認明 ⑧especially 尤其
⑨particular attention 特別注意 ⑩packing 包裝
⑪damaged 損壞 ⑫conveyance 運送

(三十) 貨物須與樣品完全相符

Gentlemen:

We received your airmailed letter of July 10, and the samples forwarded have duly come to hand.

We are pleased to find that the articles appear to be of excellent quality, and the prices and the terms of payment are also quite satisfactory to us.

We have now pleasure in sending you an order for the Chinese Toys as per the particulars enclosed. Kindly note that the goods are to be delivered in exact accordance with the samples. If this first trial order turns out satisfactory and of good value, large and increasing business may result.

When he shipment is made, you may value on us for the invoice amount, at three months' date, agreeably to your terms, and at the same time advise us of it by telegraph. Kindly acknowledge this order soonest.

Yours faithfully,

註 ①samples forwarded 送來的樣品 ②have duly come to hand已經妥收
③toys 玩具 ④as per 如同
⑤in exact accordance with 完全符合
⑥turn out 成爲 ⑦agreeably 符合
⑧advise by telegraph 電報通知
⑨at three months' date 三個月期付款

(三十一) 函購婦女用品

Gentlemen:

Please send me the following items which I have seen advertised in today's "China Post"

1. Maderite Woman's Dress in yellow
   Size 42, Style C $6.00
2. Ladies' Korona Handbags in
   black suede with strap
   Style A $10.00
3. pr. Lady Louise, 4 thread
   chiffon hosiery, size 12,
   length 32½, Tropical shade,
   @$3 $12.00

Please remove all tags from these articles, and gift wrap each separately, with the exception of the stockings, which I should like to have put in two boxes (2 pair to a box). Please "charge and send"these articles by Monday of next week.

Sincerely yours,

註 ①China Post 中國郵報 ②suede 小牛皮
③strap 皮帶 ④Louise 女襪
⑤chiffon 花邊 ⑥hosiery 襪類
⑦tropical shade 夏裝的色度 ⑧tag 價目標籤
⑨gift wrap 禮物包裝 ⑩separately 公開
⑪exception 除外 ⑫stockings 女用長襪
⑬charge 記帳

ORDER BLANK

ORDER NO. ____________

DATE ____________

Peterson Automatics Corp.
200 Fourth Avenue
New York 10., N. Y.

Gentlemen:

Enclosed please find $550.00 ________ in the United States currency, to be applied on the following order.

Quantity	Description	Price Each	Total Price
2	New "A. S." Vender	$275.00	
	Net Price		$550.00
	100 Brass Checks with machine		Free
	1 Box containing 1,000 rolls		
	(Additional rolls in cases containing 2,000 rolls can be had at $10.00 per case)		
	No extra charge for careful Export Packing. This price includes freight from factory to any American seaport.		

Address ____________

Special marks if any ____________

Consular invoice to read ____________

Remarks ____________

Signature ____________

註 ①Order Blank 空白定單　②vender 販賣機

③brass 銅 ④check 牌
⑤roll 捲 ⑥extra charge 額外費用
⑦consular invoice 領事證明書 ⑧remarks 附註

PURCHASE ORDER FORM

JOHNSON OPTICAL INSTRUMENT CO., INC.

Fouth Avenue, New York 10, N. Y.

To: Messrs. Wan & Co., Ltd. Ship: at once
1 Nanking Road
Taipei, Taiwan

Please ship the following merchandise:

YOUR CATALOG NUMBER	QUANTITY	DESCRIPTION	PRICE
5112	350 only	Magnifiers 1×2 as per sample submitted	50¢ ea.

Delivery: Prompt.
Payment: Against Letter of Credit to be established in accordance with our agreement.

Acknowledge this order promptly upon receipt and advise definite delivery date. This order is non-valid unless signed by member of our firm or the authorized employee.

Yours very truly,
JOHNSON OPTICAL INSTRUMENT CO., INC.
(Signed)

註　①Purchase Order Form 購貨單　②optical instrument 光學器材
③magnifier 放大鏡　④definite delivery date 確實交貨日期
⑤non-valid 無效　⑥authorized employee 被授權的職員

第五節　交貨書信

ADVICE OF SHIPMENT LETTERS

提　示

(1) 貨物運出後，應立即函告買方。或先用電報簡單通知，再以航空快信詳告。

(2) 倘發生困難無法如期交貨，貴方應函告買方，說明困難所在，請其諒解，並商補救辦法。

(3) 如交貨延期，以致與信用狀內的規定不符時，貴方應請買方轉請銀行修改信用狀。

(4) 函末應表示謝意，並希望以後源源賜顧。

(一) 船位不敷請同意分期裝運

Dear Sirs:

Re: Lauan Plywood under Your Order PO/723

We acknowledge with thanks the receipt of an Irrevocable Letter of Credit No. P/D-5130 you have established in our favor through Hang Seng Bank for the captioned order.

Apropos of delivery of this order, we confirm the despatch of the following telegram to you on September 14th:

"PO723 DUETO SPACE LIMITATION ONLY 700000 SQUARE FEET LOADED PLEASE AMEND LC ALLOWING PARTIAL SHIPMENT LETTER FOLLOWS"

Much to our regret, we could not deliver the total quantity by one shipment for lack of space of the carrying vessel. Since this is an incident unforeseen by us, we hereby request you to amend the L/C by deleting the special clause reading "partial shipment not allowed". Your compliance with our request will be greatly appreciated

For the remaining 300,000 square feet, we will ship them by the first boat available after receipt of your amendment to the L/C.

Faithfully yours,

註 ①Lauan plywood 柳安合板　②in our favor 爲了我們的
③Apropos of 關於　④captioned 事由上所說的
⑤despatch 發出
⑥dueto (due to) 由於（在電報中可併爲一字）
⑦space limitation 船位所限　⑧amend 修改
⑨partial shipment 分批交運　⑩letter follows 函詳
⑪much to our regret 十分遺憾　⑫lack 缺少
⑬incident 偶發事件　⑭unforeseen 未能預見
⑮delete 刪除　⑯special clause 特別條款
⑰compliance 照辦

（二）無船裝運請同意延期交貨

Re: Your Order for Cotton & Blended Yarns

Dear Sirs:

We have been notified by Bank of Taiwan that a letter of credit has been opened by you in our favor through The Hamburg Bank with them as the advising bank. From BOT'S notification, we know that the deadline of shipment for the first batch of yarns was set on September 25, 1981.

Since receipt of the Bank's advice, we have been making great efforts to book space to effect shipment on time; but, much to our regret, we were told by the shipping companies contacted that there would be no vessels sailing to Hamburg direct before September 25th. As the L/C stipulates that transhipment is not allowed, we hereby request you to have the L/C amended by extending the shipment date and its validity to October 10 and 20, 1989 respectively.

Since this is an urgent matter, please amend the L/C by cable. Your compliance with our requests will be highly appreciated.

Faithfully yours,

註 ①have been notified 被通知 ②Bank of Taiwan 臺灣銀行
③advising bank 通知銀行 ④BOT's notification 臺灣銀行的通知
⑤deadline 限期 ⑥first batch 第一批
⑦yarn 紗 ⑧book space 位定
⑨on time 準時 ⑩contacted 接觸過的
⑪stipulate 規定 ⑫transhipment 換船裝運
⑬extending 延長 ⑭validity 有效期
⑮by cable 用電報

(三) 貨已交海元輪運出

Re: Canned Crab Meat

Dear Sirs:

We take pleasure to inform you that we have effected shipment of forty-five (45) cases of the captioned food per S. S. "Hai Yuan" covered by the letter of credit No. TW-507 you have opened in our favor through the Bank of America. The vessel sailed from Keelung on September 26th.

Enclosed you will find the following shipping documents each in quadruplicate:

Commercial Invoice No. MTF-57

Packing List No. P-MTF-73

Insurance Policy No. 152 issued by China Insurance Company, Ltd.

Consular Invoice issued by American Consulate in Taipei

Bill of Lading No. OSKE-7

Survey Report No. 65 issued by the International Superintendence, Inc.

We hope that the goods will arrive at New York in perfect condition and that you will find all these documents in order.

Faithfully yours,

註 ①have effected shipment 已實行交運

②S. S. "Hai Yuan" 海元輪 (S. S.＝steamship)

③Bank of America 美國商業銀行 ④shipping documents 裝船文件

⑤quadruplicate 一式四份 ⑥commercial invoice 商業發票

⑦packing list 裝箱單 ⑧insurance policy 保險單

⑨consular invoice 領事證明書 ⑩consulate 領事館

⑪bill of lading 提貨單 ⑫survey report 檢驗報告書

⑬in order 無誤

(四) 因火災延期交貨請修改信用狀

Re: "Jet Size" Sizing Stuff

Dear Sirs:

We acknowledge with thanks the receipt of your irrevocable letter of credit opened through Bank of America for 1,000 kgs. each of Nos. 55 and 100 of the subject stuff

According to stipulations of the L/C, the total quantity should be shipped not later than September 20, and no partial shipments are allowed. Although we wished to comply with such conditions, we, however, could not do so because of a recent fire at our factory which has completely destroyed all our stocks. Therefore, we hereby request you to have both the date of shipment and the validity of the L/C extended not later than October 10 and 15, respectively.

As fire is a force majeure, we are sure that you would comply with our request accordingly.

Faithfully yours,

註 ①sizing stuff 塗料 ②subject stuff 事由上所說之用料
③stipulation 規定 ④partial shipment 分批交運
⑤comply 遵照 ⑥recent fire 最近的火災
⑦completely destroyed 全毀 ⑧validity 有效期
⑨respectively 各別地 ⑩force majeure 不可抗力
⑪accordingly 如前所述

(五) 函輪船公司預定船位

Dear Sirs:

Under a sales contract with our client in West Germany, we are to ship 100 bales of cotton yarn, weighing 400 pounds each, to Hamburg direct before September 25th.

We are writing to book shipping space with you and shall appreciate your quoting us the lowest freight rate therefor at your early convenience.

As the L/C of our client stipulates that transhipment is not allowed, therefore you must quote us for a vessel sailing from Keelung to Hamburg direct before the deadline as mentioned herein.

Your early quotation will be appreciated.

Yours Faithfully,

註 ①sales contract 售貨合同　②client 顧客
③bale 包　④booking shipping space 定船位
⑤appreciate 感謝　⑥freight rate 貨物運費率
⑦deadline 限期　⑧quotation 報價

(六) 期前無船至西德可否延到十月裝運

Dear Sirs:

We thank you for your letter of September 6, 1989 asking us to quote ocean freights to Hamburg for 100 bales of cotton yarn.

Upon receipt of your enquiry, we have contacted a large number of shipping companies as well as shipping agents and regret to inform you that there is no vessel, either Conference or Non-Conference, that sails from Keelung to Hamburg direct prior to September 25. Therefore, we are unable to quote you as requested.

For your information, there is M. V. "Singapore" of the P & O Steam Navigation Co., Ltd., which is scheduled to sail from Keelung to Hamburg direct on or about October 5. Please let us know if you would wish to ship these 100 bales by that boat.

Faithfully yours,

註 ①shipping companies 輪船公司 ②shipping agents 輪船公司代理行
③vessel of Conference or Non-Conference運費同盟或非運費同盟的船隻
④prior to 早於 ⑤M. V. (motor vessel) 輪船

(七) 覆告同意延期請速預定船位

Dear Sirs:

We are in receipt of your letter of September 10, 1989 and thank you for the information given therein.

Our client in Hamburg has agreed to extend the shipment date of their L/C to October 10, and, therefore, we can ship the 100 bales of cotton yarn per M. V. "Singapore" scheduled to sail from Keelung about October 5. You are hereby requested to book shipping space therefor and let us know the ocean freights thereof as early as possible.

Your prompt attention to this matter and early confirmation will be appreciated.

Faithfully yours,

註 ①information 消息　　②have agreed 已同意
③extend 延長　　④confirmation 證實

(八)西德在臺無領事館請刪除領事證明書一項

Re: Shipment of 100B/S Cotton Yarn

Dear Sirs:

Further to our letter of September 17 advising the booking of M. V. "Singapore" to deliver the captioned yarn, we are now requesting you to amend the letter of credit you have opened in our favor by deleting the requirement to submit Consular Invoice along with other shipping documents.

As there is no West German Consulate here in Taiwan, nor anyone representing it, we are unable to obtain the document required. Therefore, you are requested to amend the L/C accordingly.

Since the date of shipment is imminent, please have the L/C amended by cable. Your compliance with our request is hereby solicited.

Faithfully yours,

註 ①further to our letter of September 17 續九月十七日敝函
②deleting 刪除　　③consular invoice 領事證明書
④West German Consulate 西德領事館
⑤represent 代表　　⑥imminent 即將來到
⑦compliance 照辦　　⑧is solicited 被請求

(九) 函告暫定船期

Re: <u>Cotton and Blended Yarns</u>

Dear Sirs:

We confirm with thanks the receipt of your letter of August 11, 1989 and are pleased to receive the order you have placed with us.

For the 300,000 pounds of 40's/1 (100% cotton on cones) and 200,000 pounds of 45's/1 blended yarn (65% Polyester/35% Cotton in hanks), we are giving you hereunder our tentative shipping schedules:

40's/1 300,000 lbs. 100% Cotton 100,000 lbs. per month,
November, December, 1981
and January 1982

45's/1 200,000 lbs. Blended Yarn 50,000 lbs. per month,
October, November,
December, 1981 and
January, 1982.

Please arrange to open L/C's in accordance with the preceding schedules. As agreed upon between us, all L/C's should be opened one month prior to each and every monthly shipment.

We thank you again for this order and assure you that we will spare no efforts to process this order to satisfy you in every respect.

Faithfully yours,

註 ①blended yarn 混合紗　②cone 圓錐體
③polyester 人造纖維　④hank 捲
⑤tentative 暫定　⑥shipping schedule 船期

⑦in accordance with 按照　⑧preceding schedules 上面所定的船期

⑨spare no effort 不遺餘力　⑩process 辦理

⑪in every respect 在各方面

(十) 貨已裝船運往漢堡

Re: Your Order for Canned Asparagus

Dear Sirs:

We acknowledge with thanks the receipt of your Letter of Credit No. A-1058 opened in our favor through the Hamburg Bank for the captioned order and confirm the despatch of the following telegram to you on September 18th:

"LC A1058 SHIPPED 20MTS EACH TIPCUTS CENTERCUTS ENDCUTS SS EURYPAETES LEFT KEELUNG 17TH ETA HAMBURG OCTOBER 15TH AIRMAILING DOCUMENTS"

To facilitate your taking delivery, we are submitting and enclosing the following shipping documents, each in duplicate:

Commercial Invoice No. GTC-55

Packing List No. P-GTC-55

Certificate of Insurance No. MI-3251 issued by Tai Ping Insurance Co., Ltd.

Bill of Lading No. KEHA/25 issued by the Ocean Lines, Ltd.

Special Customs Form No. 323

We trust that you will find all these documents in order.

Faithfully yours,

註 ①tipcuts 前段　②centercuts 中段
③endcuts 末段　④ETA (estimated time of arrival) 預計抵達日期
⑤facilitate 方便　⑥duplicate 一式二份
⑦in order 無誤

(十一) 罐頭洋菇已運出一部份

Re: Your Order No. GICL/503 for Canned Mushroom

Gentlemen:

We are pleased to inform you that we have received your Letter of Credit No. C-15203 opened through The First National City Bank, N. Y. and that we have effected shipment of the first four items under the subject order per S. S. "Oriental Despatcher" which sailed from Keelung on September 16th.

To facilitate your taking delivery, we are enclosing three copies each of the following shipping documents:

Commercial Invoice No. TFC-37
Packing List No. P-TFC-37
Bill of Lading No. KENY-5 issued by the Oriental Lines

Certificate of Origin issued and certified by Taipei Chamber of Commerce.

Certificate of Insurance No. TP-21 issued by Tai Ping Insurance Co., Ltd.

Consular Invoice issued by American Consulate, Taipei
Special Customs Forms Nos. 5515 and 5213

For this shipment, we despatched a telegram to you on September 16th, reading:

"LC 15203 SHIPPED FIRST FOUR ITEMS TOTALING 7000 CASES SS ORIENTAL DESPATCHER 16TH ETA NEWYORK OCTOBER 20TH"

We trust you will receive the goods in perfect condition.

Yours faithfully,

註 ①First National City Bank 花旗銀行　②taking delivery 提貨
③Certificate of Origin 產地證明書　④in perfect condition完好情況

(十二) 無船直達印尼請同意在香港轉船

Re: SWC Sugar

Dear Sirs:

We are in receipt of your Letter of Credit No. B-503 for amount of US$1,050,000.00 opened in our favor through The Hongkong & Shanghai Banking Corporation to cover the FAS cost of 10,000 metric tons of the subject sugar and thank you for it.

With regard to shipment, we regret very much to inform you that, despite strenuous efforts having been made by us, we are still unable to book space of a vessel sailing to Jakarta direct. The shipping companies here told us that, for the time being, there is no regular boat sailing between ports in Taiwan and Jakarta. Therefore, it is very difficult, if not impossible, for us to ship these 10,000 metric tons of sugar to Jakarta direct.

In view of the difficult situation faced by us, you are requested to amend the L/C to allow transhipment of the goods in Hongkong where arrangements can easily be made for transhipment. Please be assured that we will ship the goods to Hongkong right upon receipt of the L/C amendment. Since this is something beyond our control, your agreement to our request and your understanding of our position will be highly appreciated.

We are anxiously awaiting the amendment to the L/C.

Faithfully yours,

註 ①SWC sugar 特級白砂糖　②in receipt of 收到
③The Hongkong & Shanghai Banking Corporation 滙豐銀行
④FAS cost 船邊交貨價格　⑤metric ton 公噸
⑥with regard to 關於　⑦regret 遺憾
⑧despite 雖然　⑨strenuous efforts 辛苦的努力
⑩book space 定位　⑪shipping companies 輪船公司
⑫for the time being 暫時　⑬regular boat 定期船
⑭in view of the difficult situation faced by us 鑒於我們面對的困難情形
⑮transhipment 轉船　⑯amendment 修改
⑰understanding 諒解　⑱anxiously 焦急地

（十三）請求加入運費同盟為會員

Taiwan Representative

Dear Sirs:

In order to avail ourselves of the good and swift services of vessels of the Conference lines, we are writing you to apply for membership to the Conference and shall appreciate your passing our application to your Hongkong Office for consideration.

For your information, we are makers of garments and, during the past two years, have exported large quantity of our products to the United States, South and East Africa, as well as to Europe. The total tonnage of our garments exported amounts approximately to 2,500–3,000 measurement tons per annum, which is substantial enough to warrant us to join the Conference as a member.

Your prompt screening of our application and approval will be appreciated.

Faithfully yours,

註 ①representative 代表　②avail ourselves 利用我們
③swift 迅速　④conference lines 運費同盟所屬輪船公司
⑤membership 會員資格　⑥application 申請書
⑦garments 成衣　⑧measurement ton 以尺碼折算的噸位
⑨per annum 每年　⑩substantial 大量
⑪warrant 認可　⑫screening 審查

(十四) 覆告申請加入運費同盟已獲核准

Dear Sirs:

We acknowledge the receipt of your letter of September 10, 1989 applying for membership to the Far Eastern Freight Conference and are pleased to inform you that your application has been approved by our Hongkong Office.

Enclosed please find the blank forms of application and other relevant documents all in triplicate. Please fill them out and return them to us as early as practicable.

Your observance of the required formalities is hereby requested.

Faithfully yours,

註 ①Far East Freight Conference 遠東運費同盟

②blank form 空白表格 ③relevant 有關的

④fill out 填好 ⑤observance 照辦

⑥formality 例行手續

(十五) 貨物因合約關係不能交運費同盟船隻裝運

Dear Sirs:

We are very much pleased to be admitted as a member of the Conference and assure you that we will from now on abide by all its regulations and rules.

Now, we are facing a quandary: our buyers in Sweden insisted that the garments we are to deliver to them be shipped by vessels of the Marchessini Line, which is a Non-Conference line because they have contractual obligations with that particular line to have all the goods shipped to them by their vessels. Meanwhile, the L/C our buyers have opened to us also calls for the goods to be shipped by boats of the Marchessini Lines.

Since this case is beyond our control, we can do nothing but comply with our buyers' request. As this is an exceptional case, we report it to you just for information and reference.

Faithfully yours,

註 ①be admitted 被允加入 ②abide by 遵守

③regulations 條例 ④rules 規章

⑤quandary 困惑 ⑥Sweden 瑞典

⑦Non-Conference line 非運費同盟所屬的輪船公司

⑧contractual obligation 合同義務 ⑨call for 要求

⑩this case is beyond our control 本案非我們所能控制

⑪comply with buyers' request 遵照買方的請求

⑫exceptional case 例外

(十六) 通知原棉即將裝船啓運

Dear Sirs:

SHIPMENT ADVICE

According to advice we have received from our Suppliers, shipment of your esteemed order of 100 M/T's of South Brazilian Raw Cotton, 1-1/32", Type: BSIK, as per our Contract No. CO-361 dated September 2, 1981 will be effected per S. S. "DITTE SKOU", scheduled to sail from South Brazilian Port on or about October 15, 1989.

The above information is given to you merely for your guidance and hence without any engagement on our part.

Thanking you for your kind attention.

Faithfully yours,

註 ①Shipping Advice 裝運通知書 ②esteemed order 貴方定單

③South Brazilian Raw Cotton 南巴西原棉

④as per 如同 ⑤merely 僅僅地

⑥guidance 指引 ⑦engagement 約束

⑧on our part 在我們方面

（十七）函告棉花已運到基隆

Dear Sirs:

INWARD CARGO NOTICE

We have pleasure in advising you that the 450 bales of raw cotton from Mombasa of which you are the notify party, arrived Keelung on October 15, 1989.

Please present the bill of lading to our Keelung Agents, Messrs. Thai Ho Co., Ltd., in exchange for the necessary delivery order.

We take this opportunity to draw your kind attention that, as stated on the relative bill of lading, the carrier's responsibility for the cargo ceases immediately the cargo leaves the ship's tackle and thereafter all risk and expense involved in the responsibility of the cargo. The cargo must also pay for coolie hire, landing, storing and delivery charges, and, if necessary, lighterage. According to Keelung Harbour Bureau regulations, the storage is payable commencing five days after the arrival of the vessel.

Customs regulations require that consignees pay duty within 15 days of the arrival of the vessel, failing which, overtime charge will be levied up to a limit of 3 months. If the delivery of the cargo is not taken at the end of the 3 months, it will be treated as smuggled goods and may be confiscated.

Faithfully yours,

註 ①Inward Cargo Notice 進口貨物通知書

②coolie hire 搬運伕費 ③bill of lading 提貨單

④delivery order 交貨通知單 ⑤draw your attention 提請注意

⑥carrier's responsibility 運送人的責任
⑦cease 中止 ⑧tackle 轆轤
⑨landing charge 上岸費 ⑩storing charge 存倉費
⑪delivery charge 交貨費 ⑫lighterage 駁船費
⑬Keelung Harbour Bureau 基隆港務局
⑭storage 倉租 ⑮commencing 開始
⑯customs regulations 海關章程 ⑰consignee 收貨人
⑱duty 關稅 ⑲failing which 如不照辦
⑳overtime charge 逾時費 ㉑be levied 被徵收
㉒up to 直到 ㉓treat 作爲
㉔smuggled goods 走私貨物 ㉕confiscate 充公

(十八) 同意開罐器延期交貨

Re: Our Order No. 70/813 for 5,000 Tin Openers

Dear Sirs:

We refer to your letter dated August 19, 1989 wherein you informed us that you are unable to deliver the subject Tin Openers out of stock because of very heavy demands.

We understand the difficult position you are in and, therefore, agree to extend the delivery date of the subject order for another two weeks. By that time, we are sure your suppliers would have replenished your stocks.

This letter serves as our confirmation of our order on the revised delivery schedule.

Faithfully yours,

註 ①tin opener 開罐器　　②out of stock 從存貨中
③replenish 補充　　④revised delivery schedule 修改後的交貨期

(十九) 玩具及雜貨數日內卽可啓運

Gentlemen:

We thank you for your esteemed letter of July 30, with an order for the Chinese toys and sundries, which shall have our best attention.

We have arranged for an early delivery and hope the shipment will be made during the next few days, when we shall not fail to advise you of it by wire.

We sincerely thank you for your valued order, and we hope that this may be a forerunner of many others.

Yours very truly,

註 ①esteemed letter 尊函　　②sundries 雜貨
③by wire 用電報　　④valued order 寶貴的定單
⑤forerunner 先驅

(二十) 雜貨已交海南輪運出

Dear Sirs:

We have the pleasure to enclose the invoice and B/L for 50 cases sundry goods shipped by M. S. "Hai Nan" in execution of your order of 20th April. We trust the goods will reach you in due course and give you entire satisfaction.

As instructed, we have drawn upon you for the net amount of £50-0-0 at sight through the Hua Nan Commercial Bank, Ltd., and would ask you to give our draft your kind protection as usual.

We trust this purchase will bring you a good profit and result in your further orders.

Yours very truly,

註 ①B/L (bill of lading) 提貨單

②M. S. "Hai Nan" 海南輪 (M. S. =Motor Ship)

③execution 執行　　④in due course 到相當時候

⑤as instructed 如尊囑所示　　⑥drawn upon you 向你開出滙票

⑦at sight 見票卽付

⑧Hua Nan Commercial Bank 華南商業銀行

⑨draft 滙票　　⑩protection 保障照付

(二十一) 玩具已經運出

Gentlemen:

We have the pleasure to inform you that the mechanical toys and sundries you kindly ordered were duly shipped today by M. S. "Hai Ping" sailing tomorrow, August 24, from Keelung. We trust that the goods will reach you safely and give you every satisfaction.

As desired, we have drawn upon you for the invoice amount, $145, 500, at three months' date, through the First Commercial Bank, and would ask you to give our draft your kind protection. Herewith we hand you the invoice and the bill of lading, both of which we trust you will find in order.

We thank you again for your valued order, and trust we may be favored with further orders in near future.

Yours very truly,

註 ①mechanical toy 機械玩具 ②sundries 雜貨

③M. S. "Hai Ping" 海平輪 ④as desired 符合尊願

⑤at three months' date三個月期 ⑥bill of lading 提貨單

(二十二) 豌豆二十噸已由海安輪裝運

Dear Sirs:

Your kind order for 20 tons of peas, which was given us on September 5, was shipped to you by M. S. "Hai An," sailing from Keelung today. We have telegraphed to you to that effect, and the copy of the wire is enclosed herein.

The copies of the invoice and other shipping documents are enclosed in this letter. According to your L/C, we have drawn upon you through the Chang Hua Commercial Bank, at 3m/s for the invoice, that is $45,000.

We trust that the peas will reach you safe, and hope to be favoured with your further increasing commands.

註 ①pea 豌豆 ②M. S. "Hai An" 海安輪

②to that effect 為了那件事 ④wire 電報

⑤shipping documents 運貨所需文件 ⑥according to 按照

⑦Chang Hua Commercial Bank 彰化商業銀行

⑧3m/s 三個月期 ⑨commands 吩咐

(二十三)水泥卽運並催欠款

Dear Sirs:

We thank you for your order of the May 2 for 500 sacks of cement, and are pleased so state that arrangements have been made with our forwarding agents for an early delivery.

In accordance with your request, the goods will be insured against all risks, the premium and the forwarding charges being payable to Messrs. Tait & Co. Every effort will be further given to select such cement as suitable to your needs, and we trust that our efforts to assist you will be successful. As soon as the shipment is ready, we will advise you by wire.

By the way, from the accompanying statement you will observe that your indebtedness to us for charges, up to this date amounts to $5,000 for which sum we should be grateful to receive a cheque at your earliest convenience.

Yours faithfully,

註 ①sack 袋　②cement 水泥
③state 說　④arrangement 安排
⑤forwarding agents 轉運公司　⑥delivery 交貨
⑦in accordance with 按照　⑧be insured against all risks 已保全險
⑨premium 保險費　⑩forwarding charges 運送費
⑪select 選擇　⑫suitable to your need 合於尊需
⑬by the way 再說(說話者忽然想到一件與本題無關的事，用此來改變話題的口頭語)
⑭accompanying statement 隨附的帳單

⑮indebtedness 欠款 ⑯up to this date 到今天爲止

⑰cheque 支票（同 check）

（二十四）原料缺乏無法立即交貨

Dear Sirs:

We thank you for your order of the October 18, but regret that an unforeseen scarcity of raw materials has rendered us unable to execute your kind order immediately. Every effort, however, is being made to deliver the necessary materials this week, and we should in consequence be able to resume production to deliver the goods by the end of this month.

We trust that the delay will not cause you any serious inconvenience, and hope that the delivery within the time stated will meet your requirement.

Yours faithfully,

註 ①unforeseen 未預見的 ②scarcity 短缺

③raw material 原料 ④render 使得

⑤execute 執行 ⑥in consequence 結果

⑦resume 恢復 ⑧delay 延誤

⑨cause you any serious inconvenience 引起你任何嚴重的不便

⑩within the time stated 在上述時間之內

⑪meet your requirement 應付你的需要

（二十五）襯衫料祇能先供應二千碼

Gentlemen:

We thank you for your order of May 6 for 5,000 yds. of cotton shirtings, but regret to advise you of our inability to supply more than 2,000 yds. immediately. The demand for this material has been enormously large, and our strenuous effort could not cope with the rush of orders.

Should you be good enough to wait until the end of this month, we shall do our utmost to comply with your kind order, because we shall be able to make considerable amount of the shirtings in our factory during the time.

We trust that you will appreciate our difficulty, and manage to put off your requirement till the end of this month.

Yours faithfully,

註 ①yds. (yards) 碼　②cotton shirting 棉質襯衫料
③inability 不能　④supply 供應
⑤immediately 立即地　⑥strenuous effort 辛苦的努力
⑦enormously 非常地　⑧cope with 應付
⑨rush 湧到
⑩should you be good enough to wait 如蒙惠予等待
⑪do our utmost 盡我們的最大力量　⑫considerable amount 大宗
⑬appreciate 察諒　⑭put off 延遲

（二十六）待信用狀到即將糖交船裝運

Re: SWC Sugar

Dear Sirs:

Your letter of August 14, 1989 placing an order with us for 10,000 metric tons of the captioned sugar has been received and we thank you for it.

We are now awaiting the arrival of your letter of credit and will ask shipping companies to book space as soon as your L/C arrives. In this regard, we will spare no efforts to comply with your request to have these 10,000 M/T's shipped as early as practicable. Moreover, we wish to draw your attention to the fact that the price concluded between us is on an FAS Kaohsiung basis; therefore, our responsibility ceases when the goods are delivered along the ship side. As previously agreed, we will ask the shipping company to collect freights from you in due course.

We look forward to receiving your L/C at the earliest.

Faithfully yours,

註 ①SWC sugar 特級白砂糖 ②metric ton 公噸
③shipping companies 輪船公司 ④book space 定船位
⑤L/C 信用狀 ⑥in this regard 關於此事
⑦spare no effort 不遺餘力 ⑧comply with 遵照
⑨price concluded 已定價格 ⑩FAS 船邊交貨
⑪responsibility 責任 ⑫cease 停止
⑬as previously agreed 如以前所約定
⑭collect freight 收取運費用 ⑮in due course 到相當時候

(二十七) 機器零件須三個月後始能交貨

Re: Spinning Machine Parts

Dear Sirs:

We are pleased to be informed in your letter of August 24, 1989 that you have already applied for import license and letter of credit for this order. As advised in our letter to you dated July 20, we can only effect shipment of the parts under this order four months after order confirmation. Since you urgently need these parts, we will try our utmost to deliver them within three instead of four months after your confirming the order. This is a special favour we give to our old client and we are sure you would appreciate it.

We are awaiting the arrival of your letter of credit in due course.

Faithfully yours,

註 ①spinning machine parts 紡紗機零件 ②order confirmation 定貨確定
③urgently 緊急地 ④try our utmost 盡我們最大力量
⑤favor 優待 ⑥client 顧客
⑦appreciate 贊許

(二十八) 因火災衣料被焚不能立即交貨

Dear Sirs:

We thank you for your order of August 5, 1981 for 200 suits in the sizes specified and trust that your requirements are not so urgent as to render an immediate delivery essential.

Owing to a recent fire, our stocks are considerably destroyed and the high-grade materials necessary for the suits are not now available. If, however, you require the suits for immediate sale, we can make them in materials nearly as good; but as the original material is likely to be extremely popular this season, we would advise you to wait about three weeks, by which time we shall be in a position to supply it.

We should be grateful to receive your opinion very soon.

Faithfully yours,

註 ①suit 一套衣服 ②sizes specified 規定的尺寸
③render 使得 ④essential 重要
⑤owing to a recent fire 因爲最近的火災
⑥stocks 存貨 ⑦considerably destroyed 大量損毁
⑧high-grade 高級 ⑨material 衣料
⑩available 可以有 ⑪nearly as good 差不多同様好
⑫extremely popular 十分流行 ⑬in a position 能夠
⑭grateful 感激 ⑮opinion 意見
⑯fire 火災

(二十九)覆請用替代品趕製衣服應付急需

Subj: Our Order for 200 Suits

Dear Sirs:

We have received your letter of August 17, 1989 and appreciate very much your advice in connection with the subject order.

Much to our regret, we learned from your letter under reply that a recent fire has considerably destroyed your stocks of woolen cloth, thus making you unable to deliver the suits we ordered on time. Although it might be advisable for us to wait another three weeks for the delivery of suits made of high-grade materials, they are now being urgently needed to meet the brisk demands of our customers. Therefore, we request you to start processing this order right away with the substitute materials which, according to your opinion, are nearly as good as the original.

Your prompt confirmation is anxiously awaited.

Faithfully yours,

註 ①appreciate 感謝 ②advice 忠告
③in connection with 關於 ④woolen cloth 毛絨布
⑤on time 準時 ⑥brisk demand 旺盛的需要
⑦processing this order 辦理此批定貨
⑧right away 立即 ⑨substitute 替代品
⑩original 原貨 ⑪anxiously 焦急地

第六節 交付貨款書信

PAYMENT LETTERS

提 示

(1) 國際貿易的一般付款方式是請銀行開信用狀。
(2) 定購貨物後，買方應函告賣方，說明已由某某銀行於某月某日開出信用狀。

(3) 國內交易可用支票或滙款交付貨價。

(4) 收到貨款後應函告買方，並表謝意。

(5) 催收欠款函，措詞必須客氣。縱然到了非法律解決不可時，仍應委婉說明出於萬不得已，希望訟案不致發生。

(6) 連續催收函，中間應隔的時間，視情形而定，通常為十天或半個月。

(一) 函告已開信用狀

Gentlemen:

This will acknowledge receipt of your cable reading as follows:

FIFTY CASES CANNED CRAB MEAT SHIPPED S/S HAI YUAN FEBRUARY 25 STOP OPEN L/C IMMEDIATELY

and thank you for your information that the 50 cases canned crab meat have been shipped by S/S "Hai Yuan" on February 25.

Upon your instructions, we have advised our bank, the First Commercial Bank today to open by cable Letter of Credit for our above order for 50 cases canned provisions in the amount of $1,500.00.

We trust that we will receive confirmation in the next few days of the shipment.

Very truly yours,

註 ①cable 電報 ②canned crab meat 罐頭蟹肉
③stop 句點（電報中用此表示此句已完）
④information 消息 ⑤upon your instruction 遵照指示
⑥have advised 已通知 ⑦provisions 食品

(二) 貨已收到支票附上

Gentlemen:

We are pleased to inform you that the goods ordered on October 12 have been received, and found entirely satisfactory both in quality and prices.

In settlement of the amount of your invoice, less 3 per cent discount, we enclose a cheque, value $350, 000. We should be pleased to have your acknowledgement in due course.

Yours very truly,

註 ①entirely satisfactory 完全滿意 ②settlement 結算
③invoice 發票 ④3 per cent discount 百分之三折扣
⑤acknowledgement 認收 ⑥in due course 在相當時候

(三) 函告貨款已如數收到

Gentlemen:

We are in receipt of your letter of October 5 enclosing a cheque, value $350, 000, which has been passed to your credit with thanks.

We trust we shall soon have the pleasure of receiving a repetition of your orders.

Yours faithfully,

註 ①in receipt of 收到 ②passed to your credit 收入尊戶
③repetition 繼續而來

（四）收到支票盼再定貨

Gentlemen:

We acknowledge receipt of your letter of May 18, enclosing a check, value $20,100 in payment of your account. We have placed the amount to your credit with thanks. The formal receipt of which you will find enclosed herein.

We assure you of our best attention being paid to your esteemed orders at all times, and we await your further commands.

Yours very truly,

註 ①in payment of your account 惠付的貴帳　②assure 保證
③esteemed order 貴戶定單　④further commands 將來的吩咐

（五）寄月終結帳單請付款

Gentlemen:

Herewith we send you the monthly statement up to and including April 30, and, as the amount this month is only $5,000, we suppose you will prefer to send us a cheque, as this is too small to draw a bill for.

Yours very truly,

註 ①herewith 附在此處　②monthly statement 月結單
③up to 截至　④including 包括
⑤suppose 猜想　⑥prefer 較願
⑦draw a bill 開滙票

(六) 款已收到附上收條

Dear Sirs:

Your letter of March 30 is to hand, containing a cheque for $56, 500, which we have placed to your credit and for which we thank you, Enclosed is the receipt as requested for the amount.

Encl. a/s　　Yours very truly,

註 ①to hand 收到　②contain 內有

③have placed to your credit 已收入尊帳

(附件)

RECEIPT

Taipei, March 31, 1989

No. 100

RECEIVED from Mr. C. Y. Kuo

the sum of Fifty-Six Thousand Five Hundred Dollars only.

By Cheque.

THE CHINA TRADING CO., LTD.

$56, 500. 00　*Per* *Sam Wang* (Revenue Stamp)

註 ①receipt 收據　②revenue stamp 印花稅票

(七) 催收逾期帳款(第一次)

Dear Sirs:

Your account shows an overdue balance of $567.90 which was

payable on February 20.

We should be grateful to receive your cheque at your convenience.

Yours truly,

註 ①overdue balance 過期懸帳　②payable 應付
③grateful 感激

(八) 催收逾期帳款(第二次)

Dear Sirs:

We refer to our letter of March 15 in which we drew your attention to the overdue balance of our January statement of $567.90.

We must assume that this account has escaped your attention and we should be glad if you would look into this matter without delay.

Yours truly,

註 ①refer 查考　②draw your attention 提請注意
③assume 假定　④escaped your attention 未蒙注意
⑤look into 查明　⑥without delay 勿延

(九) 催收逾期帳款(第三次)

Dear Sirs:

Though we have reminded you in our letter of March 15 of the overdue balance of your account, we have so far not received your check.

We are unable to keep this balance any longer and must request payment of $567.90 by the end of March at the latest.

Yours truly,

註 ①remind 提醒　②so far 到現在為止
③longer 長久　④at the latest 最遲

(十) 催收逾期帳款(第四次)

Dear Sirs:

We have asked you for settlement of the overdue amount of $567.90 in our letters of March 15 and 20. We are surprised that we have not even had a reply to our letters.

No item of the account is in dispute and we must now insist on an immediate settlement.

Please note that we shall have to hand this matter to our solicitors if your check is not received by the end of March.

We need not tell you how much we should regret such a step after long and friendly connection with your firm and we hope that you will help us to avoid it by giving this matter your immediate attention.

Yours truly,

註 ①settlement 結帳　②surprised 驚異
③dispute 爭論　④insist 堅持
⑤have to hand this matter to 不得不將此事交給
⑥solicitor 律師　⑦step 步驟
⑧long and friendly connection 長久而友好的關係

第七節 要求賠償書信

CLAIM LETTERS

提示

(1) 要求賠償函內，應詳述貨物損失經過，以及為什麼應由貴方負責的理由。

(2) 如貨物品質不符，必須全部退貨時，可請貴方更換貨品或將貨款退還。如雙方有往來帳戶，可請將所退貨款收入買方的帳戶。

(3) 為顧全信用及希望維持雙方友好關係起見，如果數目不太大，貴方最好認賠了事。

(一) 運印尼砂糖一部份損失請求賠償

Dear Sirs:

We have just received a letter from our client in Indonesia advising that about 10% or, to be exact, 9520 gunny bags have been found broken with a total loss of approximately 4, 500 kgs. of sugar. According to the Surveyor's Landing Report, one copy of which is enclosed for your reference, the breakage of the bags is due mainly to the fragility of the gunny bags. The Surveyor has further found that the bags are woven with jute of very inferior quality. Therefore, your Corporation, instead of either the shipping company or the insurer, should be held responsible for the loss of the goods.

In view of the preceding, you are requested to compensate our client in Jakarta for the total loss of sugar at the FAS value of US$105.00 per metric ton.

Your early settlement of this case will be appreciated.

Faithfully yours,

註 ①client 顧客　②advising 通知
③exact 正確　④gunny bag 蔴袋
⑤approximately 約計
⑥Surveyor's Landing Report 公證人的貨物起岸報告書
⑦for your reference 供備參考　⑧breakage 破碎
⑨due mainly to 主要地由於　⑩fragility 易破
⑪jute 黃蔴　⑫inferior quality 次等質料
⑬shipping company 輪船公司　⑭insurer 保險公司
⑮responsible 負責　⑯in view of the preceding 鑒於上述情形
⑰compensate 賠償　⑱settlement 解決

(二) 臺糖公司同意賠償

Dear Sirs:

Your letter of November 5, 1981 together with one copy of Landing Report has reached us. Much to our regret, considerable loss has been inflicted owing to the fragile gunny bags used for packing the sugar.

Upon receipt of your letter, we have given this matter our immediate attention and instructed our laboratory to give a strict test of the durability of the gunny bags. We are much surprised that the findings of our laboratory happended to be the same as stated in the Landing Report: the gunny bags have been found not strong enough to bear weight of 100 kgs.

Based on the findings, we agree to compensate your client in Jakarta for the total loss and enclose one Sola Draft No. 103/YS503 for US$472.50 issued by Bank of Taiwan to pay therefor, please pass the draft on to your client and let us have their formal receipt in due course.

Truly yours,

註 ①inflict 使受 ②fragile 易破
③instruct 吩咐 ④laboratory 試驗室
⑤strict 嚴格 ⑥durability 耐用力
⑦surprise 驚奇 ⑧finding 查出的結果
⑨Sola Draft 單張滙票(無副張) ⑩formal receipt 正式收據
⑪in due course 在相當時候

(三) 運來水泥品質不合但願折價購買

Dear Sirs:

We regret to bring to your notice certain grave irregularities in connection with our order No. 128/48 for 7,500 sacks of Portland Cement.

Your representative, Mr. Chen, plainly stated that the cement to be supplied was entirely impervious to the action of water. Our analyst, however, states that it is quite useless for the purpose intended; further, we must point out that Mr. Chen was informed definitely of the use to which it would be put.

You will recognize that we are accordingly in a position to repudiate the whole contract, but such a drastic step would be most unwelcome to us. It is clear, however, that we are entitled to some compensation, and we should be glad to hear of the allowance you are prepared to make to meet the case.

Yours faithfully,

註 ①grave 嚴重的　②irregularities 不符合
③in connection with 關於　④sack 袋
⑤portland cement 波特蘭水泥　⑥representative 代表
⑦plainly 明白地　⑧impervious 不能透過的
⑨analyst 化驗師　⑩definitely 確實地
⑪recognize 承認　⑫repudiate 拒絕
⑬drastic step 激烈的步驟　⑭are entitled to 應得
⑮compensation 補償　⑯allowance 折扣

（四）臺泥公司覆允按七折計價

Dear Sirs:

We must ask you to accept our apologies for our error in supplying you with a grade of cement which we now find to be of a quality very inferior to that offered by Mr. Chen and ordered by you.

We are quite prepared to have this cement returned, carriage forward, but as you seem to be in a position to use it, we should make an allowance of 30 per cent if you definitely decided to retain the whole consignment.

Disciplinary action has been taken against those responsible for the error and a repetition of the trouble is most unlikely. We should accordingly be grateful if you would still honour us with your confidence, and we hope that you will entrust us with the supply of the original quality required.

Yours faithfully,

註 ①apology 道歉　②error 錯誤
③inferior 較次等的　④retain 保留
⑤consignment 運出的貨　⑥disciplinary 懲戒的
⑦repetition 重複　⑧unlikely 不大可能
⑨confidence 信任　⑩entrust 信任

(五) 手錶不符請告如何處理

Dear Sirs:

The consignment covering our order AW/1899 arrived last week. We were sure disappointed to find that the watches have only luminous hands but no luminous dials as shown in the catalog.

We can only accept these watches with a substantial allowance and are holding them at your disposal pending your reply.

Yours truly,

註 ①disappointed 失望　②luminous 發光
③hand 時針　④dial 時計面
⑤catalog 目錄　⑥substantial 很大
⑦at your disposal 任君處置　⑧pending 懸待

(六) 貨物件數短少請速補運

Dear Sirs:

Your consignment of clocks arrived today and has been found correct with the exception of Model A of which 10 were ordered and invoiced while the case contained only 4.

Please look into the matter without delay and send the missing goods by air freight as we can accept them only if they arrive before the end of the month.

Yours truly,

註 ①look into the matter 調査此事 ② without delay 勿延
③missing 失落的 ④air freight 航空貨運

(七) 覆已將短少件數補寄

Dear Sirs:

Your letter of July 4 has crossed ours of June 30 in which we informed you that the mistake in our consignment had been noticed and that the six Model A clocks had been shipped by air freight free of charge.

We apologize once more for this most regrettable mistake and have taken measures to prevent a recurrence of similar errors in future.

Yours truly,

註 ①your letter has crossed ours 尊函與敝函中途錯過
②mistake 錯誤 ③free of charge 免費

④take measures 採取措施 ⑤prevent 防止
⑥recurrence 再發生 ⑦similar 相似

(八) 運來女傘品質太差且多破損

Dear Sirs:

We have received from you 10 dozen ladies' silk umbrellas as per your invoice of May 12.

However, upon opening up these umbrellas, we find that they are not at all satisfactory. We find that some of them have the handles off, the frames are rusty and five have the ribs broken. The cloth is defective, seeming to have holes in it.

Under the circumstances, we thought it better to return these umbrellas to you as per enclosed return bill.

Yours very truly,

註 ①umbrella 傘 ②as per 如同
③handle 柄 ④frame 架
⑤rusty 上銹 ⑥rib 傘骨
⑦defective 有缺點 ⑧return bill 退貨單

(九) 玩具不佳唯有退貨

Gentlemen:

We received 4 dozen Chinese toys as ordered, but found some had split wheels, while others did not move at all, owing to the poor spring in the workmanship.

We are returning these to you, for which kindly allow credit.

Yours very truly,

註 ①split wheel 裂輪　②spring 彈簧
③workmanship 技藝　④allow credit 准予收入帳帳

(十) 來貨不合市場需要

Dear Sirs:

We have returned by air parcel the 5 dozen rubber shoes received from you last week.

These do not suit our trade and we have returned same.

We would appreciate receiving a Credit Note covering the goods returned at your earliest convenience.

Thank you.

Yours very truly,

註 ①air parcel 空運包裹　②rubber shoe 橡膠鞋
③suit 合於　④same 原物
⑤credit note 入帳通知單

(十一) 內衣質劣決定退回

Dear Sirs:

This is to inform you that we have been receiving numerous complaints in regard to the ladies' underwears which we have purchased from you recently.

As you know, we purchased these underwears from you on the understanding that they were the first quality goods.

Knowing that we were dealing with a reliable firm, we did not take the trouble to examine these underwears before sending them out. As a result, the greater part of our shipments have been returned to us by our customers who are very dissatisfied with this merchandise, terming them 'seconds', or worse.

This situation is very serious to us, as not only does this affects the sale of the underwears, but also causes our clients to lose confidence in us, jeopardizing our future business relations with them.

We enclose copies of letters received from some of these concerns who have returned the goods to us. The letters speak for themselves. The originals are on file in our office, ready for your perusal at any time.

We are returning these damaged underwears for credit. As it is more than likely further returns will be coming in to us from our customers, we reserve the right to send these back to you as they come in.

We hope and trust that in future, all shipments to us will be carefully inspected, so that this kind of embrassment and possible loss of future business, will be avoided.

We thank you for your courtesy at all times.

Very truly yours,

註 ①numerous 許多　　②complaint 責難

③underwear 內衣　　④recently 最近地

⑤understanding 默契，瞭解　　⑥reliable 可靠的

⑦firm 公司，行號　⑧touble 麻煩
⑨customer 顧客　⑩seconds 次貨
⑪worse 更壞　⑫serious 嚴重
⑬affect 影響　⑭confidence 信心
⑮jeopardize 損害　⑯The letters speak for themselves 信內已說明一切
⑰original 原件　⑱on file 存卷
⑲perusal 細閱　⑳reserve the right 保留權利
㉑inspect 檢查　㉒embrassment 困擾
㉓avoid 避免　㉔courtesy 禮遇

（十二）枱燈損壞請求賠償

Dear Sirs:

The goods you shipped per S. S. "Hai Nan" on 14th last month arrived here yesterday.

On examination, we have found that many of the desk lamps are severely damaged, though the cases themselves show no trace of damage.

Considering this damage was due to the rough handling by the steamship company, we claimed on them for recovery of the loss; but an investigation made by the surveyor has revealed the fact that the damage is attributable to improper packing. For further particulars, we refer you to the surveyor's report enclosed.

We are, therefore, compelled to claim on you to compensate us for the loss, $175, which we have sustained by the damage to the goods.

We trust that you will be kind enough to accept this claim and deduct the sum claimed from the amount of your next invoice to us.

Yours faithfully,

註 ①severely 嚴重地　②trace 痕跡
③damage 損壞　④rough handling 粗魯的處理
⑤recovery 賠償　⑥investigation 調查
⑦surveyor 公正人　⑧reveal 透露
⑨attributable to 歸因於　⑩improper packing 不合宜的包裝
⑪for further particulars 至於進一步的詳情
⑫are compelled 被迫　⑬compensate 補償
⑭sustained 遭受　⑮deduct 減少

（十三）覆告提議解決枱燈賠償辦法

Dear Sirs:

We have received your letter of 15th July, informing us that the desk lamps we shipped to you arrived damaged on account of imperfectness of our packing.

This is the first time that we have received such a complaint from consignees, although we have always been shipping such goods with similar packing as we shipped the goods to you.

We are convinced that the present damage was due to extraordinary circumstances under which they were transported to you. We are therefore not responsible for the damage; but as we think it would not be fair to have you bear the loss alone, we suggest that the loss shall be divided between both of us, to which we hope you will agree.

Yours faithfully,

註 ①desk lamp 檯燈　②imperfectness 不完善
③consignees 收貨人　④convinced 相信

⑤extraordinary 非常的；意外的⑥circumstances 情況
⑦fair 公平 ⑧bear 擔負
⑨agree 同意

（十四）紅露酒一箱損壞請速處理

Gentlemen:

We regret to inform you that one of cases of your consignment arrived in a badly damaged condition. It is PK/49 containing 150 doz. "Red Dew" wine. The lid was broken and the case with its contents crushed. It looks as if some very heavy cargo has fallen on it. We have examined the boxes and find that about 30 dozens are in unsaleable condition.

We pointed out the damage when taking delivery and have endorsed our receipt "one case severely damaged".

We assume that you are taking the matter up as the insurance had been effected by you.

Truly yours,

註 ①consignment 來貨 ②Red Dew Wine 紅露酒
③lid 蓋 ④crush 壓破
⑤severely 嚴重地 ⑥cargo 貨物
⑦unsaleable 不能售出 ⑧delivery 提貨
⑨endorse 背書，加註 ⑩assume 假定
⑪take this matter up 提出交涉

（十五）覆告酒已另寄十二打

Dear Sirs:

We regret to see from your letter of October 2 that one case of our shipment arrived in a badly damaged condition. As the goods were packed with the greatest care we can only conclude as you say that the case has been stored or handled carelessly.

We have reported that matter to our insurance company. Will you please hold the damaged goods at our disposal; we will give you credit when the matter has been settled.

Meanwhile we have dispatched 12 doz. "Kaoliang" wine to replace the damaged ones and we trust that this will be in accordance with your wishes.

We shall write to you again as soon as we have heard from the insurance company.

Sincerely yours,

註 ①conclude 認爲　②at your disposal 任你處置
③credit 入帳　④settle 解決
⑤meanwhile 同時　⑥despatch 寄出
⑦Kaoliang Wine 高粱酒　⑧replace 代替
⑨ones 那些（指損壞的酒）　⑩in accordance with 按照

第八節　貨物保險書信

INSURANCE LETTERS

提　示

(1) 函請保險公司投保時，必須將所保貨物名稱、數量、總價、賠款付款地點、保險種類等，一一詳細列明。

(2) 保險公司多有印好之空白投保書，逐項填明即可。

(3) 不用空白投保書，用普通信函投保也可以。

(4) 保險公司覆函時應將保單隨附，並告知保費。

(一) 函保險公司申請保險

Dear Sirs:

Kindly issue a Marine Insurance Policy in the name of The National Trading Co., Man Yee Building, Taipei.

Claims payable at London.

Amount of Insurance required: £4,000.

Against the risk of All Risk & War Risk

Mark, Quantity & Description of Merchandise:

Leather Shoes. One hundred(100) dozen packed in five cases, @ 20 dozen each.

Marks:

NTCC
TAIPEI
No. 1/5

Voyage from Taipei to London per S. S. "Wing On" sailing on or about September 10, 1989.

Yours faithfully,

註 ①issue 簽發 ②Marine Insurance Policy 水險單
③in the name of 戶名 ④claim 賠償
⑤payable 付給 ⑥amount 金額
⑦all risk 全險 ⑧war risk 兵險
⑨mark 標誌 ⑩description 內容
⑪merchandise 貨物 ⑫leather shoe 皮鞋
⑬voyage 航程 ⑭S. S. Wing On 永安號輪船

(二) 向保險公司投保水險

Gentlemen:

We should be obliged by your insuring against all risks $60,000, value of 100 cases of Canned Provisions, marked ◇SS◇ No. 1/100, shipped at Keelung, on board M. S. "Hai Yuan," sailing for New York on Nov. 9. Please send us the policy, together with a note for the charges.

Yours very truly,

註 ①obliged 感謝 ②canned provisions 罐頭食品
③on board 在船上 ④policy 保險單
⑤charges 費用

(三) 覆已遵照保險

Gentlemen:

Pursuant to your instructions dated Nov. 8, we have insured your shipment of 100 cases of Canned Provisions, marked ◇ No. 1/100, shipped at Keelung on board M. S. "Hai Yuan" sailing for New York on Nov. 9, as per the Policy enclosed. We hand you herewith our account for $4, 200, which amount please pass to our account.

Yours very truly,

註 ①pursuant to 遵從　②instruction 指示
③as per 如同　④pass to our account 收入敝帳戶

(四) 茶葉受損請付賠款

Dear Sirs:

Re: S. S. "Silver Star"

We are holders of the Policy No. 31404 issued by your Hong Kong agents on 400 cases tea valued at HK$25, 000.00. During the voyage the ship encountered heavy weather, and in consequence 150 cases were damaged by sea water. We enclose the certificate of survey, also the Policy, which is against marine risk W. A.

Kindly adjust the claim and send us a cheque in settlement at your earliest convenience.

Yours truly,

註 ①holders 持有人　②agent 代理
③voyage 航程　④encounter 遭遇到
⑤heavy weather 惡劣天氣　⑥in consequence 結果
⑦certificate of survey 檢驗證明書　⑧W. A. 水漬險
⑨adjust 調整　⑩settlement 解決

(五) 覆附茶葉賠款支票

Dear Sirs:

We are in receipt of your letter of May 26, regarding 150 cases of damaged tea per M. S. Silver Star. We have pleasure in enclosing a cheque for HK$8,000 in settlement of your claim. We shall be glad to have your receipt in due course.

Yours very truly,

註 ①in receipt of 收到　②per 由
③pleasure 愉快　④in due course 在相當時候

(六) 手帕受損索賠

Gentlemen:

Re: Claim on Handkerchiefs
per S/S MEXICO MARU
Under policy No. BN/5611

Referring to the captioned claim, of which we sent you a preliminary notice on Oct. 3, 1989, we hereby file a detailed claim for the amount of $376.26 as per statement attached.

Kindly acknowledge receipt of this letter and settle the claim as soon as possible.

Yours truly,

Encl.: Statement of claim
Certificate of Insurance

Invoice

Survey Report

Packing List

Letter of Carrier (Copy)

註 ①handkerchief 手帕 ②preliminary 初步的

③file a detailed claim 提出一詳細要求 ④statement attached 附表

⑤survey report 檢驗報告書 ⑥packing list 裝箱單

⑦carrier 承運人

(附件)

STATEMENT OF CLAIM

Vessel: Mexico Maru

Voyage: Osaka/New York

Date of Arrival: Oct. 2, 1989

Shipper: C. Itoh & Co., Osaka

Commodity: 95 cartons handkerchiefs

Invoice: CI—334, valued at $13,923.00

Insurance Certificate: BN/5611, covering All Risks

Insured Amount: $22,877.00

Nature of Claim: Sea Water Damage, non-delivery

PARTICULARS OF CLAIM

Insured Amount: $22,877.00

Insurable Value:

F. O. B. Value ·················· $13,923.00

Freight & Insurance ·············· 1,282.50

Handling ···························· 54.34

Trucking............................ 85.00

$15, 344. 84

@ 1. 4906

Sea Water Damage:

1120 doz. #5813

@ $0. 35/doz.

Insurable Value $392. 00

Damaged at 33. 33% $130. 65

Non-delivery:

300 doz. #5813

@ $0. 35/doz.

Insurable value.. 105. 00

Total $235. 65

@ 1. 4906 .... $351. 26

Survey fee 25. 00

Amount claimed.. $376. 26

內容要點

(I) 本索賠計算書包括二部分，前一部分記載與索賠有關的一般事項，後一部分記載索賠項目及計算。

(II) 索賠項目包括①海水損害(1120打) 計美金130. 65元，②遺失 (300打) 計美金105元，合計美金235. 65元。

(III) 由於每一美金之貨物投保 1. 4906元，故 235. 65 元的貨物應賠 351. 26 元。此數再加公證費用 25 元，使索賠金額增至美金 376. 26元。

有關保險的幾點說明

保險的種類很多，有火險、水險、人壽險、意外險、竊盜險、信用險等等，但和國際貿易發生最密切關係的却是水險 (Marine Insurance)，因為目前世界各地的貨運，仍以海運為最普通。

水險方面，一般分為三種：（一）全部損失險(Total Loss Only)簡稱T. L. O.；（二）平安險（Free From Particlular Average）簡稱F. P. A.；（三）水漬險（With Particular Average）簡稱 W. P. A. 或 W. A. 。

全部損失險是指貨物遭遇到全部損失時方予賠償，否則，只是部份的損失時，投保人是不能獲得賠償的。這一種保險，投保人所享受的權益實在太少了，所以投這種保險實不多見。

平安險，它的原名是 Free From Particular Average, "Average" 這個字在保險的方面的意義是「海損」。所以，平安險這個名字是譯得不恰當的，應稱為：「不保單獨海損險」才適當。不過平安險這名字，經過長時期沿用下來，已無人把它更正了。平安險的保險範圍是比較全部損失險為大的。在平安險的條款中有訂明：船舶在航運中遭遇觸礁、碰撞、焚燬等均得到賠償；但如貨物僅因海難（暴風雨等）而遭受到單獨海損，在平安險的條款下是得不到賠償的。

水漬險，它的原名是 With Particular Average，正確的譯名是：保單獨海損險。不過水漬險這個名字，因沿用過久，也無人更正了。水漬險的保險範圍比平安險更擴大了。即是說：除了平安險的範圍保險之外，其他如大風浪、船漏水等海難而引起的損失都可獲得賠償。

然而，因有些情形而引起貨物的損失，仍不受到水漬險條款所擔保的，如：

Theft, Pilferage and Non-Delivery (T. P. N. D.) 偷竊及短交險、Stain（油漬）、Breakage（破爛）、Fresh and Rain Water（淡水損壞）、Hook Hole（鈎孔）Acid（酸蝕）、Leakage(洩漏)、Damaged By Other Cargo（受其他貨物影響而受損）……等。

上述那些是叫做附加保險，那些附加保險可選出需要的那一項來附加投保的。不過現在的商人覺得太麻煩了，如果覺得水漬險仍不足保障安全時，便索性投保一切險（All Risk）即是說連上述那些附加保險都包括了。所以一切險這方式現在用得很普遍了。

最後還有一項是兵險（War Risk），兵險就是戰爭險，即是說因戰爭而引起不論捕捉、擄押、扣留或由武器所致的損失均屬兵險所保範圍。

兵險和水險不同的，所以水險中的一切險條款仍不包括兵險在內的。因此我們投保時除了在水險中選擇一種外，仍須另投保兵險的。

第九節　寄售貨物書信

CONSIGNMENT LETTERS

提　示

(1) 寄售貨物不採用信用狀付款方式，因此被委託代售的外國公司行號，必須信用可靠，否則難免有收不到貨款的風險。

(2) 代售的佣金多少，如果雙方早已商定，在委託的信內可無須說明。

(3) 售出全部或一部份貨物後，責方應開列帳單，函告貨主。

（一）請代售貨物

Gentlemen:

We are in receipt of your letter of the April 6 and in order to try your market, we have shipped to your consignment per M. S. "Hai Yua'„u the goods specified in the annexed invoice, of which we enclose Bill of Lading. In conformity with your request, we have this day drawn on your friends Messrs. Richardson Wang & Co., New York, against the invoice amount, $1,000 at 30d/s.

We hope that the goods will reach you in good condition realizing profitable prices. We await to receive your advice.

Yours very truly,

註 ①market 市場　②consignment 寄售貨物
③specified 列明　④annexed invoice 隨附的發票
⑤Bill of Lading 提貨單　⑥in conformity with 符合
⑦request 請求　⑧this day 今天
⑨have drawn on your friends 已向貴友開出滙票
⑩30d/s 三十天期　⑪realize 實現
⑫advice 通知

（二）寄售方糖

Gentlemen:

In conformity with the request of your New York correspondents, we have much pleasure in making you a consignment of cube sugar.

Enclosed we now hand you the Bill of Lading and Invoice for the twenty cases best cube sugar, marked ◇ 1/20, per S.S. "Hai An" to sail for your port tomorrow. We trust that all may be found in order, and that the cases may reach you safely, and in good condition. Commending the sugar to your care for disposal to the best advantage in our mutual interests, we wait for the Account Sales and returns in due course.

Yours truly,

註 ①correspondent 聯行 ②cube sugar 方糖
③port 港口 ④in order 無誤
⑤commend 委託 ⑥disposal 處置
⑦mutual interest 雙方利益 ⑧returns 報告書

(三) 請代試銷耶誕飾品

Gentlemen:

Our good friends, Messrs. Charles Chang & Co., of Taipei, with whom, we understand, you have extensive dealings, have suggested your name to us, as being in a good position to handle our consignment to advantage

Acting solely upon their advice, we are sending you a trial consignment of 10 cases Christmas Fancy Goods, | C. C. | 1/10, per M. S. "Hai Ping" leaving Keelung on August 10, which we trust will reach you in good time and order, and realize profitable prices.

Enclosed please find the Consular Invoice, Bill of Lading, and Insurance Policy for the goods, also ordinary Commercial Invoice, amounting to $9,000. Terms as usual, viz., 5%.

We shall be glad to learn that our proposal is satisfactory to you, and we hope that we may have a long and profitable connection.

Yours very truly,

註 ①extensive dealings 廣泛的交易 ②suggest 建議

③solely 僅僅地，完全地　　④advice 忠告
⑤fancy goods 新奇的貨品　　⑥consular invoice 領事證明書
⑦terms 條件　　⑧as usual 照舊
⑨viz (namely) 就是　　⑩proposal 建議
⑪connection 聯繫

(四) 覆允代銷耶誕飾品

Gentlemen:

We are in receipt of your letter of August 8, advising us the despatch of a consignment of X'mas Fancy Goods per S. S. "Hai Ping." Upon arrival of the goods, we shall use our best endeavours in disposing of same to your satisfaction.

At the moment we have many inquiries for the line of your consignment, and we feel certain that your goods will realize profitable business. We thank you for your giving us a trial, and we assure you of our best services at all times.

Yours very truly,

註 ①receipt 收到　　②despatch 寄出
③X'mas=Christmas 耶誕節　　④endeavour 努力
⑤in disposing of same 處理該批貨物
⑥at the moment 現在　　⑦inquiries 詢問
⑧realize 實現　　⑨trial 嘗試
⑩assure 保證

(五) 函告耶誕飾品已代售罄並獲厚利

Gentlemen:

We have the pleasure of informing you that the X'mas Fancy Goods received by S. S. "Hai Ping" were profitably disposed of.

You will see in the Account Sales, which we enclose, that all the goods were cleared at better prices than we expected. We are pleased that in this first instance we have been enabled to give you such a satisfactory report.

The net proceeds due to you we hold at your disposal, and we should be pleased to know what we could do with them. We await to receive further favors from you.

Yours very truly,

註 ①were profitably disposed of 已獲利地處置了

②Account Sales 售貨帳　③cleared 售完

④expected 預料　⑤first instance 第一次

⑥net proceeds 淨得之款　⑦due to you 應給你的

⑧at your disposal 任憑處置

(六) 鳳梨五十箱已代售完

Gentlemen:

We have been favored with your letter of Feb. 16 and feel very much obliged to our worthy friends, Messrs. Charles Chang & Co., of Chicago for their kind recommendation of us to you.

The fifty cases canned pineapples consigned to us by S. S. "Hai An" have already arrived and been safely landed, and we have disposed of them, prior to the landing of several other cargoes forwarded from your city. We are confident that you will approve of this course of action, when we assure you that the market has fallen fully ten per cent since we last sold our canned fruits.

We have the pleasure now to enclose you the account sales of them; the net proceeds of which amounting to $12,500, we shall hold at your disposal.

Yours very truly,

註 ①obliged 感謝 ②worthy friends 貴友
③recommendation 推薦 ④canned pineapple 罐頭鳳梨
⑤safely landed 平安起岸 ⑥prior to 在前
⑦confident 有信心 ⑧approve 同意
⑨market has fallen 市價已下跌

有關寄售貨物的說明

國際貿易的正常方法，是由售方透過銀行開發信用狀，於裝船後取得貨款。但有時為了試銷新產品，或試拓新市場，商人願意將貨運往他地，委託一家商行，代為推銷。以售得貨款的百分之幾，作為該行代銷佣金。這種貿易方式，當然帶些風險，不如信用狀方式的迅速、安全和可靠。

第十節 求職與應聘書信

APPLICATION FOR JOB LETTERS

提 示

(1) 報上所登招聘或事求人廣告可分為二類，一為說明聘請機關或公司名稱的 (Open Advertisement)。另一種是僅說某機關或某公司或郵箱號數，不具真名的 (Blind Advertisement)。

(2) 如果廣告內沒有要應徵人列舉必須具備的詳細條件，那麼，應徵函祇消將學歷經歷簡單說明卽可。

(3) 應徵及求職函內的語氣要不卑不亢，在謙遜中不失自己的身份。不可以乞憐態度求對方同情。要點是表明你的學歷經歷和專長，希望你的才能，可以適合對方的需要。換句話說，要以對方的利益為重點，如此方才能受到對方的尊重。

(4) 如果久久得不到回音，應徵人或求職人不妨再去信詢問。一般說來，無論成與不成，對方總會有回覆的。倘使對方請你去面談(Interview)，那至少有一半的希望了。

(5) 去信的好壞，對謀職的關係很大，因為對方對你的第一個印象就是從信裏得來。所以這封信寫時要特別當心，無論內容外觀，務必力求精美。

(6) 求職和應徵函，宜用手寫，當然打字也可以。

(一) 應徵速記打字員工作

Gentlemen:

Your advertisement for a secretary in this morning's "China Post" interests me greatly. I should like to be considered an applicant for this position.

As it is difficult for me to know from the advertisement the exact requirements you wish of the applicant, I shall list my qualifications briefly:

I am a graduate of the Taipei City Commercial High School (with honors). Here I completed all the work of the secretarial course in stenography and typewriting. I also attended the Caldwell Business School and further improved my shorthand and typewriting. I can now take shorthand dictation at a rate of 125 words a minute with facility.

I should be glad to submit my references to you at your request. May I call for an interview at a time most convenient for you? You may write to me at the address above or call me at 681-5321.

Yours very truly,

註 ①advertisement 廣告　②secretary 秘書
③China Post 中國郵報　④applicant 應徵人
⑤position 職位　⑥exact 正確
⑦qualification 資歷　⑧graduate 畢業生
⑨course 科目　⑩stenography 速記
⑪shorthand 速記　⑫dictation 聽寫
⑬facility 熟練　⑭submit 呈送
⑮reference 諮詢人　⑯interview 面談
⑰call 打電話

(二) 應徵簿記員職務

Dear Sirs:

Your advertisement in this morning's Central Daily News for a bookkeeper was shown me by a friend of mine.

I was graduated last March from Taiwan Commercial College with A-1 rating, and I have a letter of recommendation from the President of the College.

While I have no actual business experience, I believe that I can satisfactorily fill a position where no expert work is required.

I should be glad to come to you on trial, leaving you to determine the value of my work so that you may fix my salary after a week or so.

Mr. S. Wang, Director of the First Commercial Bank, and Mr. T. Shen, President of the Taiwan Trading Co., Ltd., have given me permi-ssion to use their names as reference as to my reputation for honesty and willingness to work.

I hope I shall have the pleasure of a favourable answer.

Yours respectfully,

註 ①Central Daily News 中央日報 ②bookkeeper 簿記員
③A-1 rating 最佳成績 ④letter of recommendation 推薦函
⑤experience 經驗 ⑥expert 專門
⑦on trial 試用 ⑧determine 決定
⑨fix salary 敍薪 ⑩permission 允許
⑪reputation 聲譽 ⑫willingness 願意

(三) 應徵秘書工作

Dear Sirs:

Please consider me as an applicant for the position of a secretary in your Bank, which is advertised in this morning's China Times.

I am twenty years of age and a graduate of the Min Chuan Commercial College, where I have completed the four-year course. Besides, I have had two years' training in stenography and typewriting, and also studied the English language, including a year of Business English. I have also completed, with a grade of A, a years' course in Business Practice, but I have had no experience in office work outside our College.

In regard to my character and general ability, I refer you to President I. M. Bao of the Min Chuan Commercial College, and Mr. O. H. Wang, Dean of the Business Administration Faculty.

I should welcome a personal interview.

Yours respectfully,

註 ①China Times 中國時報 ②Min Chuan Commercial College 銘傳商專
③have completed 已完成 ④grade of A 甲等
⑤character 品行 ⑥ability 能力
⑦personal interview 親自面談

(四) 應徵出納員職位

Gentlemen:

Having heard that the situation of a cashier in your office is vacant, I should like to offer my services for it.

I am 21 years of age, and have worked for eighteen months in a similar position for Messrs. John & Co., Taipei, to whom I can refer you as to my character and ability. I left them solely on account of their making a reduction in their establishment.

Regarding salary, I leave that with you, but feel certain that I can earn NT$8,000 dollars per month. I hope that I may have the pleasure of an interview.

Very truly yours,

註 ①cashier 出納員　②vacant 空着
③similar 相似的　④solely 僅僅的
⑤on account of 因爲　⑥reduction 裁減
⑦establishment 公司　⑧regarding salary 關於薪水

(五) 應徵簿記員工作

Dear Sirs:

I have seen your advertisement in the China Post for a book-keeper, and should be grateful if you would consider a short account of my character and capabilities.

For the last two years I have been employed as a junior book-keeper by a firm of import and export merchants where the wide field of work to be covered has given me a good all-round knowledge of accounts. To supplement my practical knowledge, I have taken an evening course in the Tam Kiang College of Commerce. I feel that you could safely entrust a set of books to my care.

You will find enclosed a testimonial from the Head Professor of the College, who has kindly offered to provide you with any further details you may require.

I trust that you will consider this application favourably and I wish to assure you that I should make every effort to be worthy of the confidence you may place in me.

Yours faithfully,

註 ①short account 短報告　②capability 能力
③junior 初級　④all-round 各方面的
⑤accounts 帳目　⑥supplement 補充
⑦practical knowledge 實用知識　⑧evening course 夜間部學科
⑨Tam Kiang College of Commerce 淡江商學院
⑩entrust 信託　⑪a set of books 一套帳簿
⑫to my care 交我處理　⑬testimonial 證件
⑭provide 供給　⑮effort 努力
⑯worthy of 不負，值得　⑰confidence 信心

（六）覆請親來面談

Dear Sir:

Replying to your application for the position of a secretary, we should be pleased if you come here at 10.00 a.m. on Monday, the 15th September should you still be disengaged. We wish to arrange matters to our mutual advantage on that occasion. We hope that you will bring a Curriculum Vitae with you.

Yours very truly,

註 ①should you be 如果你　②disengaged 未有工作
③arrange 安排　④mutual advantage 雙方利益
⑤on that occasion 關於此事　⑥curriculum vitae 履歷表

（七）詢能否來行面談

Dear Miss Chow:

Dr. David Cheng, General Manager of The First National Bank, has asked me to write to you to acknowledge your application for the post of Personal Assistant and to ask if you can come and see him on Thursday, June 8 at 4 p. m.

Will you please let me know if that suits you?

Yours sincerely,
Rebecca Wang

註 ①Personal Assistant 私人助理 ②suit 合宜

（八）覆可準時到行面談

Dear Miss Wang:

Thank you for your letter of June 22 acknowledging my application and suggesting an interview. I shall be pleased to come at 4 p. m. on June 8.

Yours sincerely,
Nancy Chow

註 ①suggest 建議　②pleased 高興

(九) 函告月薪及起薪日期

Dear Miss Chow:

I am writing to confirm the offer which we made you in our conversation yesterday of the post of Personal Assistant to our General Manager.

The salary will be $10,000 per month to start with, rising by $500 per month to $15,000. The agreement can be terminated by three months' notice on either side. You will be entitled at first to two weeks' leave of absence with full pay per year.

I should be glad if you could begin your duties here on September 1. We are looking forward to having your assistance and we hope you will be happy in your work here.

Yours sincerely,
Rebecca Wang

註 ①confirm 證實　②to start with 起始
③agreement 協定　④terminate 終止
⑤either side 任何一方　⑥entitled to 應得
⑦leave of absence with full pay 休假可領全薪
⑧begin your duties 上班　⑨assistance 幫助

(十) 覆允擔任新職

Dear Miss Wang:

I have great pleasure in accepting, on terms stated in

of June 15, the position of Personal Assistant to your General Manager. The work is exactly the kind that I have always wanted to do and I can assure you of my whole-hearted cooperation in it.

I am looking forward eagerly to my new job.

Yours sincerely,

Nancy Chow

註 ①accepting 接受 ②terms 條件
③General Manager 總經理 ④whole-hearted 全心全力的
⑤cooperation 合作 ⑥eagerly 急切地

(十一) 覆告應徵職位已由他人充任

Dear Sir:

We thank you for your letter applying for the post of assistant to the Director of this firm. Your application has been carefully considered but the post has been awarded to another applicant. We were, however, favorably impressed by your qualifications and have filed your application in this office. If there is a future opening in the firm for which we think you would be suited, we shall certainly let you know.

Yours truly,

註 ①assistant to the Director 董事助理 ②awarded 給予
③applicant 申請人 ④favorably impressed 印象甚佳
⑤file 存卷 ⑥opening 空缺

(十二) 因病請假

Dear Sir:

Unfortunately due to the attack of influenza, I was taken to Taipei Country Hospital yesterday.

Under the circumstance, I have to take sick leave for a fortnight beginning October 1. If everything goes well, I hope I could return to work on October 15.

Your approval will be appreciated.

Yours sincerely,

註 ①unfortunately 不幸地　②attack 罹患，襲擊
③influenza 感冒　④Country Hospital 宏恩醫院
⑤circumstance 情況　⑥have to 不得不
⑦take leave 請假　*⑧sick leave 病假
⑨fortnight 二星期　⑩approval 批准
⑪appreciated 感謝

* 除 Sick Leave 外，尚有其他多種假，例如:

1. casual leave 事假
2. official leave 公假
3. maternity leave 產假
4. marital leave 婚假
5. compensatory leave 補假
6. home leave 在國外工作的人回國休假
7. leave without pay 留職停薪
8. absent without leave 曠職
9. French leave 不辭而別，開溜
10. annual leave 每年規定的定期休假

（十三）因病呈請辭職

Dear Sir:

It is with much reluctance and regret that I must ask to be released from my position of Chief Auditor of your Bank.

As you are aware, for the past few weeks poor health has prevented me from carrying out my duties and responsibilities in a manner satisfactory to myself.

Please accept my resignation to take effect at the end of June 1989.

Sincerely yours,

註 ①reluctance 勉強，不願意 ②regret 遺憾
③be released 被解除 ④Chief Auditor 稽核長
⑤as you are aware 如你所知 ⑥poor health 健康不佳
⑦prevent 阻止 ⑧carry out 履行
⑨duties 職務 ⑩responsibility 責任
⑪resignation 辭職 ⑫take effect 生效

（十四）推薦女秘書

To Whom It May Concern:

Miss Cynthia Chang worked as a secretary in this company for three years, from 1987 to 1989. She is very competent at typing and shorthand and has fulfilled her other duties well. As her immediate supervisor for the last three years, I can state that she is a very good secretary. She is hardworking and pleasant and is well-liked by the other members of our staff. She now wants to move on to a job with more responsibility and higher pay, for which she is surely qualified.

I can give her a whole-hearted recommendation.

Very truly yours,

註 ①competent 勝任　②fulfill 達成
③immediate supervisor 直接主管　④staff 全體職員
⑤higher pay 較高薪水　⑥qualified 合格
⑦whole-hearted 全心全力的　⑧recommendation 推薦

(十五) 推薦司機

TO WHOM IT MAY CONCERN:

Mr. Chi-wu Wang was employed as my driver from March 1, 1971 to December 31, 1989, During that time Mr. Wang proved to be a very careful and efficient operator.

He has a pleasing manner and is honest and trust-worthy.

I strongly recommend him for a similar job.

Very truly yours,

註 ①"TO WHOM IT MAY CONCERN" 是不限對象的推薦信，可以用做離職證明書。
②operator 司機　③pleasing manner 態度和藹
④trust-worthy 值得信賴　⑤similar job 相似工作

履歷表（式一）

CURRICULUM VITAE

NAME: Tony Wu

AGE: 23

DATE OF BIRTH: September 1, 1952

PERMANENT DOMICILE: 20 Tai Ping Road, Taipei

PRESENT ADDRESS: Same as above

EDUCATION:

Name of School	Location	Graduated
Taipei Primary School	Taipei, Taiwan	June 1963.
Nankang Secondary School	Taipei, Taiwan	June 1969.
Soochow University	Taipei, Taiwan	Expect to graduate this coming June

EXPERIENCE: Two summers in a cotton-textile factory in Taipei as an assistant book-keeper and typist.

REFERENCE: Dr. C.Y. Cheng, Professor, Soochow University, Taipei.

註 ①curriculum vitae 履歷表 ②domicile 家，住所
③Soochow University 東吳大學

履歷表（式二）

CURRICULUM VITAE

Name: Wang Lin	Birth Date: April 14, 1944	PHOTO
Native Province & City: Ningpo Hsien Chekiang Province		
Present Address: 45, Chung Shih Road, Section 2, Taipei		
Permanent Address: Same as above	Female Single	

EDUCATION

Grade	Name of School	Location	From Month	From Year	To Month	To Year	Graduated
Primary	Municipal Kuting Primary School	Taipei		1949		1955	Graduated
Junior High	Provincial Taipei Girls Junior High School	Taipei		1955		1958	Graduated
Senior High	Provincial Taipei Girls Senior High School	Taipei		1958		1961	Graduated
College	Education Department of National Cheng Chih Universit y	Taipei		1961		1965	Graduated

EMPLOYMENT RECORD

Name of Organization	Employed From Mo.	Employed From Yr.	Employed To Mo.	Employed To Yr.	Job Title		Salary	Reason For Leaving
Private Ta Hwa Middle School		1965		1963	Teacher		NT $ 3000	Resigned
Private Tsai Hsin Middle School		1968		1971	Teacher		NT $ 4000	Resigned

Person to notify in case of emergency	Relation	Address	Tel.
Wang Sung	Father	Same as above	461-2135

Signature　Wang Lin
Date　May 30, 1973

第十一節 金融業務書信

BANKING LETTERS

提示

(1) 金融業也是商業之一，但因為它的業務有專門性，所以另列一節。實際上在金融業務信函裏，格式和一般商業信函的寫法，完全相同。

(2) 金融業所用的術語，應在寫信時特別注意。例如有擔保的貸款為“Secured Loan”，無擔保的為“Unsecured Loan”等。

(3) 顧客申請貸款，核准的易於答覆，不准的却要委婉解釋，以期諒解。縱然是公立銀行，也不能以衙門自居，一切應以企業化為經營原則。

(一) 通知開業

Dear Sirs:

We take great pleasure in announcing the opening of our Bank on July 1, 1989 at the following address:

The Chung Hwa Enterprise Bank
89 Ming Shen Road
Taipei, Taiwan
Republic of China
Telephone: 351-8362

Cable address: "CHUNGHWA" TAIPEI

Our Bank is equipped to handle all kinds of domestic and international banking business, and we are confident that our experienced and efficient staff can render you the best service to your satisfaction.

We are sure that the opening of our Bank will start our pleasant relationship with you. We shall be grateful if you will kindly extend to us your cooperation and patronage in the days to come.

Yours truly,

註 ①announce 宣布 ②opening 開幕
③cable address 電報掛號 ④equip 準備齊全
⑤handle 處理 ⑥domestic 國內
⑦international 國際 ⑧confident 有信心
⑨experienced 有經驗的 ⑩efficient 有效率的
⑪render 提供 ⑫relationship 關係
⑬grateful 感激 ⑭extend to 給予
⑮cooperation 合作 ⑯patronage 光顧
⑰in the days to come 將來

（二）請顧客開戶

Dear Mr. Neufeld:

The First Commercial Bank in Taiwan wishes to extend to you a cordial invitation to open an account with it and to make use of the comprehensive banking facilities it offers. Our Bank is conveniently located near your office, and our experienced staff will always be glad to render you prompt and efficient services.

As you are aware. our Bank, with its worldwide network of branches and correspondents, is in a position to offer you every modern banking facility.

It is sincerely hoped that the enclosed copy of our "1980 Annual Report" will prove of interest to you.

Very truly yours,

Encl.: a/s

註 ①cordial 熱誠的 ②invitation 邀請
③open an account 開戶 ④comprehensive 豐富的
⑤banking facilities 金融上之便利 ⑥render service 提供服務
⑦prompt 迅速 ⑧network 聯絡網
⑨correspondent 聯行 ⑩in a position 能
⑪annual report 年報 ⑫located 位在

(三) 謝顧客開戶(甲)

Dear Mr. Neufeld:

We thank you very much for your kind patronage in opening a Current Account with our Bank.

In order to promote the interests of our customers, we offer every facility whenever required and hope you will avail yourself of our services which are placed entirely at your disposal.

We trust that this new relationship will bring us into amiable connection, and wish you success and prosperity in the years to come.

Very truly yours,

註 ①current account 活期存款戶 ②in order to 爲了
③promote 增進 ④to avail yourself of 你自己利用
⑤at your disposal 聽君安排 ⑥amiable 友善的
⑦prosperity 繁榮

(四) 謝顧客開戶(乙)

Dear Mr. Wang:

Thank you for your letter of April 9, 1981 enclosing a check for NT$250,000 together with the documents required for your Current Account to be opened with us.

Enclosed is a Check Book opened in your name on April 10, 1989. This account has now a balance of NT$250,000 in your favor.

We are confident that we can render you the best service to your satisfaction.

Yours truly,

註 ①together with 連同 ②document 文件
③check book 支票簿 ④in your name 用尊名

(五) 同意建立往來關係

Dear Sirs:

We thank you very much for your letter of January 26, 1989 and are happy to inform you that we agree to establish a direct correspondent relationship with your bank upon receipt of your confirmation to this effect.

As requested, we wish to designate Bank of America, New York and London as our clearing banks for liquidating the transactions in US currency and Pound Sterling respectively conducted between our two banks. Therefore, each time you issue your payment order or draft, the same amount must be remitted to one of the aforesaid clearing banks, as the case may be, to the credit of our account with them under advice to us. On the other hand, when you receive our draft and or documents negotiated by us under your letter of credit, please reimburse us according to the instructions we will indicate to you.

We are enclosing one copy of our "Terms and Conditions" and "List of Authorized Signatures" for your reference. In return we hope to receive your control documents for our files. As for telegraphic test key for authenticating cable messages to be exchanged between our two banks, we would like to use yours.

We do hope this is just a beginning of a long pleasant relationship between our two institutions and we wish to assure you of our cooperation at all times.

Very truly yours,

Encl.: a/s

註 ①direct correspondent relationship 直接聯行關係

②to this effect 對於此事　③as requested 遵照囑咐

④designate 指定　⑤clearing bank 清算銀行

⑥liquidate 清算　⑦transaction 交易

⑧US currency 美國貨幣　⑨Pound Sterling 英鎊

⑩respectively 各別地　⑪conduct 進行

⑫payment order 支付命令　⑬draft 滙票

⑭remit 滙寄　⑮aforesaid 上述

⑯to the credit of our account 收入敝戶

⑰negotiate 承兌 ⑱reimburse 付還

⑲according to 按照 ⑳instruction 指示

㉑indicate 告知

(六) 願與國外銀行聯系合作

Dear Sirs:

It is our pleasure and desire to enter into business relations with your bank. To start with, we already asked Heng Shen Bank, Hong Kong to remit to you HK$50,000 being export proceeds relative to their credit No. 2203 our EBP-1869 and with which as an initial deposit please open an account with yourselves in our name under advice to us.

We are expecting more business to come and shall be glad to work closely with you.

We look forward to a relationship that will always be beneficial to both of our institutions.

Very truly yours,

註 ①desire 願望 ②export proceeds 出口所得

③relative to their credit 和他們的信用狀有關

④initial 初次 ⑤deposit 存款

⑥advice 通知 ⑦relationship 關係

⑧beneficial 有利的 ⑨institution 機構

(七) 告貸款已核准

Dear Sirs:

We are pleased to inform you that your request for a loan of NT $ 500,000 as outlined in your letter of May 7, 1989 has been approved by our Bank.

Please come to our Bank at your earliest convenience in order to prepare documents necessary for the release of funds.

Yours truly,

註 ①request for a loan 申請一項貸款 ②outlined 概述
③approve 核准 ④prepare documents 準備文件
⑤release of funds 撥款

(八) 告貸款未獲通過

Dear Sirs:

We regret to inform you that your application of June 9, 1989 for a NT $ 1,000,000 working capital loan has not been considered favorably by our Loan Committee owing to the insufficient collateral you proposed to offer.

We, however, will be pleased to reconsider your request if you are willing to increase the collateral acceptable to us.

We look forward to hearing from you soon and thank you for the interest you have shown in our Bank.

Truly yours,

註　①regret 遺憾　②application 申請
③working capital 流動資金　④Loan Committee 貸款委員會
⑤collateral 抵押品　⑥acceptable 可接受的

(九) 婉却貸款案件

Dear Sirs:

We regret to inform you that we are not in a position to grant you a loan for NT$500,000 as requested in your application dated February 2, 1981, which loan is supposed to be secured by mortgage of collaterals as listed in the attachment of the application. The reasons are given as follows:

First, our source of loanable funds are quite limited. For this reason, our loan policy is set to extending loans only to those borrowers who can give us the most collateral benefit. This means we require, as a general rule of thumb, that our loan borrower gives us business more than at least six months before we can consider making a loan to him.

Second, we, as a policy, are only making short-term loans up to six months which should be paid off upon maturity. Any loan in excess of a six-month period is beyond our consideration.

Although we are not in a position to assist you this time, we suggest that you make use of our other facilities for which we shall be pleased to assist you.

Very truly yours,

註　①to be secured 被保證　②mortgage 抵押
③attachment 附件　④loanable funds 可以貸放的資金

⑤quite limited 十分有限 ⑥collateral benefit 抵押品利益
⑦rule of thumb 從經驗所得的法則 ⑧short-term loans 短期放款
⑨upon maturity 到期 ⑩excess 超過
⑪beyond our consideration 不在我們考慮之內
⑫facility 設施，便利 ⑬assist 幫助

（十）外銷貸款有困難

Dear Sirs:

We refer to your application of March 8, 1989 for a pre-export loan in the amount of NT$350,000 in connection with your export of Christmas decorations to the United States.

After a careful review of the line of credit we granted you, we find that the facility for pre-export loan has been fully utilized. We regret indeed that we are no longer in a position to consider your loan until the present outstanding of pre-export loan has been reduced by the proceeds of your export negotiations.

We hope you will understand our position and will not hesitate to call on us if you have any questions in this regard.

Truly yours,

註 ①pre-export loan 外銷貸款 ②in connection with 關於
③decoration 裝飾品 ④review 檢討
⑤grant 給予 ⑥fully utilized 全部已被利用
⑦no longer in a position 不再能夠 ⑧consider 考慮
⑨outstanding 懸帳 ⑩proceeds 所得款項
⑪hesitate 遲疑 ⑫call on us 來找我們
⑬in this regard 關於此事

(十一) 告押滙銀行不應該扣除費用

Bank of America

New York

Dear Sirs:

Your Ref. No. 518/65

<u>Our L/C No. 813/501</u>

We refer to the enclosed photostatic copy of your debit advice for US$682 representing the draft amount for US$670 plus your advising charges and negotiation commission amounting to US$12 under the captioned credit.

Since our credit stipulates that all banking charges outside Taiwan are for the beneficiary's account, we are not in a position to pay these charges.

We shall appreciate it if you will look into this matter and reverse your entry by crediting our account with you US$12 under usual advice to us.

Your prompt attention and cooperation in this matter will be highly appreciated.

Yours truly,

Encl.: a/s

註 ①photostatic copy 影印本 ②debit advice 欠款通知書
③represent 代表 ④plus 加
⑤advising charge 通知費 ⑥negotiation commission 承兌佣金
⑦stipulate 規定 ⑧for beneficiary's account 由受益人負擔
⑨to reverse(an entry)冲轉 ⑩credit our account 收入敝帳戶

(十二) 信用狀不能付款請出口商還款

Far East Trading Co.

15 Li Shan Road

Taipei

Gentlemen:

Re: Our Negotiation EBP5124/7703
under L/C No. 1564 for
US$5,000,000 Issued by
Bank of America
New York

We refer you to our letter of February 10, in which we informed you that the documents you presented to us for negotiation under the subject credit had been unpaid by the issuing bank because of the following discrepancies:

(1). Late shipment,

(2). Credit expired, and

(3). Transhipment effected.

As you understand, we initiated a series of correspondence with the issuing bank in an attempt to have the transaction settled one way or another. However, the documents were refused by the drawees and our repeated efforts exerted towards this end have eventually turned out unsuccessful.

Since this transaction was paid against your Letter of Indemnity covering such discrepancies as mentioned above, we therefore request you to refund us the proceeds as soon as possible. Meanwhile, we are also pleased to help you with this case whenever you need us to do so.

We shall appreciate your cooperation and assistance in settling this transaction with us by the end of March 1981.

Very truly yours,

註 ①discrepancy 不符 ②credit expired 信用狀已過期

③transhipment effected 實行換船裝運

④initiate 發動 ⑤a series of correspondence 一連串的信

⑥issuing bank 開發信用狀的銀行 ⑦in an attempt 試圖

⑧one way or another 各種方法 ⑨drawee 滙票付款人

⑩repeated efforts 一再努力 ⑪eventually 終於

⑫indemnity 賠款 ⑬refund 退款

⑭meanwhile 同時 ⑮assistance 幫助

⑯settle 解決 ⑰transaction 交易

(十三) 將國外銀行拒付理由通知出口商

Gentlemen:

Our EB 103/245 for US$5,200.00
under L/C No. 711/65 of Irving Trust
Bank, San Francisco

With reference to the above-mentioned subject, we have received a letter from our correspondent stating that the draft was dishonored upon presentation to the drawees for the following reasons:

1. Packing List does not specify weight of individual bags.
2. Certificate of Origin was issued by the manufacturer instead of being issued by Ministry of Economic Affairs.
3. Stale Bill of Lading.

Please look into this matter and let us have your instructions as early as possible.

Yours very truly,

註 ①correspondent 聯行 ②draft 滙票
③dishonor 拒付 ④drawee 付款人
⑤packing list 裝箱單 ⑥specify 說明
⑦individual bag 個別袋 ⑧certificate of origin 產地證明書
⑨Ministry of Economic Affairs 經濟部
⑩stale Bill of Lading 陳舊提單 ⑪to look into 調查

(十四) 查詢信用狀款項

Dear Sirs:

Our EBP 0052/174 for US$5,000
Your L/C No. 144/30

We negotiated the draft with relative documents drawn under the captioned credit on October 20, 1981 and requested you to remit the proceeds to The Bank of America, New York for the credit of our account. However, to our regret, we have not received any credit advice to this effect from you nor from the clearing bank up to now.

Since this matter has been long outstanding, please investigate whether or not you have already remitted the proceeds as requested and inform us of the result as soon as possible.

Very truly yours,

註 ①negotiate 承兌 ②relative documents 相關文件
③remit 滙寄 ④proceeds 所得款項

⑤credit advice 收款通知書 ⑥to this effect 有關此事
⑦clearing bank 清算銀行 ⑧up to now 直到現在
⑨long outstanding懸擱已久⑩investigate 調查

(十五) 通知信用狀遺失

Gentlemen:

L/C No. KC 84231 for US$6,000
opened by Heng Shen Bank

We have been informed by The World Trading Co., Ltd., Taipei, the beneficiary of the captioned letter of credit, that the original credit instrument, which was advised by us, has been lost.

As requested by the beneficiary, we have issued a copy of the credit to replace the original one. You are therefore requested to take due note of the fact and refrain from effecting negotiation against the original which has been declared lost and which may be presented at your bank.

Please bring this notice to the attention of your branches concerned.

Thank you for your kind cooperation.

Very truly yours,

註 ①beneficiary 受益人 ②the original credit instrument 信用狀正本
③replace 代替 ④to refrain from 制止
⑤effecting negotiation 實行承兌 ⑥branches concerned 有關的分行
⑦cooperation 合作

(十六) 更改負責行員簽字

To Our Correspondents

Dear Sirs:

You are kindly requested to note the following changes in the Authorized Signature Book of our Bank effective March 1, 1989:

Head Office, Business Department

Mr. C. Y. Cheng has been promoted to Division Chief with authority to sign as a Class B Officer for and on behalf of this Bank.

Chung Shan North Road Branch

Mr. K. L. Lee has been appointed Manager with authority to sign as a Class A Officer for and on behalf of this Bank.

The specimen signatures of Mr. C. Y. Cheng and Mr. K. L. Lee are enclosed for your reference.

Please insert the above-stated signatures into the relevant section of our Authorized Signature Book. Kindly acknowledge receipt.

Yours truly,

Encl.: a/s

註 ①correspondent 聯行　②authorized signature book 授權簽字册
③effective 生效　④Division Chief 科長
⑤on behalf of 代表　⑥specimen signature 簽字樣式
⑦for your reference 供備參考　⑧insert 插入
⑨relevant 有關的　⑩kindly acknowledge receipt 此信收到後請惠告

(十七) 寄簽字樣本

Gentlemen:

We take pleasure in sending you photostatic signatures of the officers authorized to sign on behalf of this Bank. Please note the following instructions governing such signatures.

All checks, drafts, orders, bills of exchange, letters of credit, guarantees, deposit certificates, and all documents or correspondence authorizing payment of funds or delivery of securities or goods, except when signed by the President, must be signed by two authorized officers, one of whom must be a Class A Officer.

Ordinary correspondence and endorsement on such checks, drafts or any other items as are deposited for our credit or presented for collection for the credit of our accounts with local or overseas bankers shall require the signature of only one authorized officer.

Please sign and return the enclosed acknowledgement form.

Yours truly,

Encl.: a/s

註 ①photostatic 影印　②on behalf of 代表
③instructions 指示　④bill of exchange 滙票
⑤guarantee 保證書　⑥deposit certificate 存款證明書
⑦security 擔保品　⑧President 總經理
⑨endorsement 背書　⑩collection 代收; 託收
⑪local or overseas bankers 本地或海外的銀行
⑫acknowledgement form 收到後應塡的表格
⑬a/s 附件如文 (as stated 的簡寫)

(十八) 通知遺失滙票

Dear Sirs:

We regret to inform you that the following bank draft has been reported lost or stolen:

Bank Draft No. 211/7455 for US$5,000

Dated May 12, 1981

On the Bank of America, New York

To the order of Dr. Chen-yueh Cheng

You are requested not to effect payment against this draft, should it be presented to you for encashment.

Please inform your interested branches accordingly.

Thank you for your kind cooperation.

Yours truly,

註 ①bank draft 銀行滙票 ②reported lost or stolen 據報遺失或被竊

③to the order of 抬頭人 ④encashment 兌現

(十九) 核對帳目

Dear Sirs:

Our reconciliation with your statement of our account at the close of business on March 31, 1981 shows the following exceptions.

Date of Entry	Amount	Description
February 22, 1981	US$5,260	It appears that this item should have been credited

		to First Bank, Taipei instead of to our account.
February 26, 1981	US$ 8, 634. 50	The correct amount for this item should be US$ 8, 684. 50.

Please investigate and make the necessary adjustments under advice to us.

Thank you.

Very truly yours,

註 ①reconciliation 使一致（如帳上數目字）　②exceptions 不符
③investigate 調查　④adjustment 調整

（二十）詢某公司信用狀況

Dear Sirs:

We shall appreciate your providing us with an opinion as to the credit standing, respectability and financial responsibility of the following firm:

East Trading Co., Ltd.
65 Kuo Min Road
Taipei, Taiwan

Any information you care to give us will be treated as strictly private and confidential. We shall, of course, be happy to reciprocate your courtesy whenever you allow us the opportunity to do so.

Very truly yours,

註 ①provide 供給　②opinion 意見
③credit standing 信用狀況　④respectability 社會上的地位
⑤financial 財務的　⑥private and confidential 機密的
⑦reciprocate 報答　⑧courtesy 禮貌；恩惠

(二十一) 答覆徵信調查 (甲)

Dear Sirs:

In reply to your enquiry on The East Trading Co., Ltd. in your letter of July 10, 1989 we are pleased to give you the following information:

This Company is a highly respectable private company of import and export merchants, registered in 1950.

Authorized capital: NT$500,000.00 fully paid up. Favorably known to us for many years and regarded as trustworthy for their business engagements.

There are no mortgages and debentures.

The foregoing information is given in confidence and for your private use only and without any responsibility on the part of this Bank or its officials.

Yours truly,

註 ①information 資料　②highly respectable 卓著聲譽
③register 註册登記　④authorized capital 法定資本
⑤pay up 繳足 (如股份)　⑥trustworthy 值得信任的
⑦engagements 交易　⑧mortgage 抵押物
⑨debenture 債券　⑩confidence 機密
⑪any responsibility 任何責任

(二十二) 答覆徵信調查 (乙)

Gentlemen:

Asia Trading Company, Limited, 16 Min Chuan Road, Taipei, about whom you enquired in your letter of April 2, have been one of our desirable customers for the last ten years. During that time, we have opened numerous letters of credit on their behalf to Germany, to Korea and recently to the United States of America for sums, in some cases, exceeding US$100,000. Thus far, however, we have had no trouble in financing their transactions.

Mr. C. Y. Cheng, the president, is known for his financial responsibility and good moral character. Members of the Board of Directors are prominent in the financial circles of Taipei.

This company was organized in 1950 with a capital of NT$8,000,000 and it has now increased to twenty million New Taiwan dollars. Their balance sheet published in February this year indicates net profit at approximately NT$850,000.

From these facts and figures and from our own experiences, we have a very high regard for the business ability and financial responsibility of this company. We trust this information would prove of help to you. Of course, it is given in confidence and for your private use only and without any responsibility on the part of this bank or its officials.

Yours truly,

註 ①numerous 許多　②on their behalf 代表他們
③exceeding 超出　④thus far 直到現在

⑤financing 融資　　⑥moral character 道德品格

⑦members of the Board of Directors 董事會各位董事

⑧prominent 著名的　　⑨financial circles 金融界

⑩balance sheet 平準表　　⑪net profit 淨盈餘

⑫approximately 大約地　　⑬facts and figures 事實與數字

⑭high regard 重視

（二十三）送對帳單

Dear Sirs:

In response to your letter of August 5, 1989, we are pleased to enclose a Certificate of the Balance of your current account with us at the close of business on July 31, 1989.

Yours truly,

Encl.: a/s

註 ①Certificate of the Balance 結帳單

②at the close of business 營業終了時

（二十四）通知顧客支票遭退票

Dear Sir:

Check issued by Mr. L. Chen
for NT$50,000 on First Bank

We regret to inform you that the captioned check, with which we credited your account for NT$50,000 on June 8, 1989 has been dishonored and returned to us by the drawee bank with the reason of "insufficient funds". We have therefore taken the liberty of deducting the equivalent amount from your account.

We shall be obliged if you will bring us another check for the amount as soon as possible. In exchange for your new check, we shall be pleased to return the dishonored one to you.

Yours truly,

註 ①credit your account 收入貴帳戶　②dishonor 退票
③insufficient funds 存款不足　④take the liberty 自行
⑤deduct 減去　⑥equivalent 相等的
⑦in exchange 交換

(二十五) 請銀行發結帳單

Dear Sirs:

Would you please send us statements for our No. 670 and No. 671 accounts made up as at close of business on June 30. As these statements are needed for a directors' meeting arranged for next Thursday July 9, I should be obliged if you would send them to reach me not later than the morning of the 8th.

With the statements please include a certificate of balance on deposit account as at close of business on June 30, 1989.

Yours faithfully,

註 ①statement 結帳單　②directors' meeting 董事會議

(二十六) 通知顧客有超支情形請來面洽

Dear Sirs:

On a number of occasions recently your account has been overdrawn. The amount overdrawn at close of business yesterday was NT$540 and we should be glad if you would arrange for the credits necessary to clear this balance to be paid in as soon as possible.

Overdrafts are allowed to customers only by previous arrangement and as I notice that your account has recently been running on a very small balance, it occurs to me that you may wish to come to some arrangement for overdraft facilities. If so, perhaps you will call to discuss the matter. In the absence of such an arrangement I am afraid it will not be possible to honour future cheques drawn against insufficient balances.

Yours faithfully,

註 ①overdrawn 超支　②overdraft 透支
③previous arrangement 預早安排　④discuss 討論
⑤absence 沒有　⑥insufficient balance 餘額不足

(二十七) 覆已將存款補足

Dear Sirs:

Thank you for your letter of yesterday. I have today paid into my account cheques totaling NT$50,000. I realize that this leaves only a small balance to my credit and as I am likely to be faced with fairly heavy payments in the coming months, I should like to discuss arrangements for overdraft facilities.

Yours faithfully,

註 ①realize 瞭解　②leave 餘存
③fairly 相當的　④heavy payments 巨額付款

(二十八) 請求延期歸還貸款

Dear Sirs:

On May 5 last you granted me a loan of NT$500,000. This loan is due for repayment at the end of this month and I had already taken steps to prepare for this. But unfortunately, due to a fire at my warehouse a fortnight ago, I have been faced with heavy unexpected payments. Damage from the fire is thought to be about NT$1,000,000 and is fully covered by insurance, but as my claim is unlikely to be settled before the end of next month, I should be glad if the period of the loan could be extended till then.

I am sure you will realize that the fire has presented me with serious problems and that repayment of the loan before settlement of my claim could be made only with the greatest difficulty.

Yours faithfully,

註 ①due 到期　②repayment 還款
③take steps 採取步驟　④due to a fire 由於火災
⑤warehouse 貨棧　⑥fortnight 兩星期
⑦unexpected 未料到的　⑧damage 損失
⑨covered by insurance已經保險　⑩claim 要求賠償
⑪settle 解決　⑫extended till then 延到那時

(二十九) 支票遺失請銀行止付

Dear Sirs:

I am writing to confirm our telephone of this morning to ask you to stop payment of cheque No. 24572 for the sum of NT$15,000, drawn payable to the Tatung Electrical Co., Ltd.

This cheque appears to have been lost and a further cheque has now been drawn to replace it.

Please confirm receipt of this authority to stop the payment.

Yours faithfully,

註 ①stop payment 止付　②payable to 付給
③replace 替代　④authority 權力

(三十) 告遠期支票不能抵用

Dear Sirs:

In reply to your letter of yesterday, we express regret that we were not able to allow payment against your cheque No. 61134. It appears to have escaped your notice that one of the cheques paid in on August 20, the cheque drawn in your favour by Ta Hwa Co., was post-dated to August 25th and that the amount cannot be credited to your account before that date.

To honour your cheque would have created an overdraft of more than NT$1,000,000 and in the absence of previous arrangement we are afraid we could not grant credit for such a large sum.

We trust this explanation will make matters clear.

Yours faithfully,

註 ①have escaped your notice 未蒙注意

②post-dated 遠期的 ③overdraft 透支

④in the absence of previous arrangement 未有事先安排

⑤grant credit 給予貸款 ⑥explanation 解釋，說明

(三十一) 請舊客戶繼續往來

Dear Mr. Lin:

A banking account that has been inactive for any length of time is a matter of deep concern to us. I am writing personally to inquire why we have not been privileged to serve you recently.

Our improved service and constant attention to custom comfort and convenience in banking add to the reasons why I believe you will enjoy our service again.

We want to serve you. Please honor us with this privilege.

Sincerely yours,

F. D. Hu

Manager

Business Department

註 ①account 帳戶 ②inactive 停頓

③length 長 ④matter 事件

⑤concern 關切 ⑥personally 親自

⑦inquire 詢問 ⑧privileged 受愛顧

⑨recently 最近 ⑩improved 改善的

⑪constant 經常的 ⑫attention 注意

⑬custom 慣常的　⑭comfort 舒適
⑮convenience 便利　⑯add to 加上
⑰enjoy 享有　⑱privilege 優惠

(三十二)為帳目錯誤道歉

Dear Mr. Chang:

Kindly accept our apology for the notice recently sent to you in error. Your account is, as usual, in perfect order.

We will make every effort to prevent such mistakes and I am sorry that you have been inconvenienced by this oversight on our part.

Yours sincerely,

S. M. Lee

Manager, Collection Department

註 ①apology 道罪　②recently 最近
③error 錯誤　④account 帳目
⑤perfect 完美　⑥order 規律，狀況
⑦effort 努力　⑧prevent 防止
⑨mistake 錯誤　⑩inconvenienced 造成不方便
⑪oversight 失察　⑫on our part 在我們方面

第十二節 特種商業書信

SPECIAL TYPES OF BUSINESS LETTER

提 示

(1) 除了一般性的商業書信外，尚有若干特殊性質的商業書信，如交際信、訂位信、介紹信等，也很重要。

(2) 大體說來，這種函件，不宜冗長。但雖三言兩語，必須含有誠懇、明確的要點。

(3) 預訂旅館房間或機位、席次等，更要詳述所訂日期、時間、等級等，不可稍有含糊。覆信時應重述一遍，以確保無誤。

(4) 因服務不週遭顧客責難，在答覆時，應避免為自己辯護。不論是非曲直，應婉轉道歉，如此方可增進公共關係。

(5) 寫介紹信時，應說明被介紹人姓名、職位、與本人之關係，再說明介紹之目的，最後謝其協助。

(6) 退出社團，必須具函說明理由，並致謝悃。

(一) 祝賀開業

Dear Sirs:

We are glad to receive your announcement about the opening of your office in Taipei.

Please accept our best wishes on this memorable occasion and our words of congratulations for the success of your international business.

Yours truly,

註 ①announcement 宣告　②best wishes 祝福
③memorable 可紀念的　④occasion 事件
⑤international 國際的

(二) 祝賀當選商會會長

Dear Mr. Chiang:

I have just learned that you were elected President of the Taipei Chamber of Commerce, and I would like to extend my warmest congratulations to you on this fine recognition of your ability.

The Chamber of Commerce should also be congratulated on the splendid choice they have made. You have my best wishes for a successful term.

Sincerely yours,
S. K. Juin

註 ①learned 知道　②elected 被選，當選
③President 會長　④Chamber of Commerce 商會
⑤extend 給予　⑥warmest 最熱忱的
⑦congratulations 祝賀　⑧fine 妥善的
⑨recognition 認同　⑩ability 能力
⑪splendid 了不起的　⑫choice選擇
⑬best wishes 祝福　⑭term 任期

(三) 祝賀榮任分行經理

Dear Mr. Wang:

It gives me much pleasure to learn that you were appointed manager of your bank's Tainan Branch.

Please accept my most sincere congratulations on your promotion and my best wishes for your continued success in your new assignment.

Yours sincerely,

註 ①pleasure 高興，愉快 ②Tainan Branch 臺南分行
③sincere 誠懇的 ④congratulations 恭賀
⑤promotion 升級 ⑥best wishes 祝福
⑦success 成功 ⑧assignment 任務

(四) 祝升任會計主任

Dear Dick:

Hearty congratulations on your promotion to Chief Accountant of National Chemical Company. I know how happy you and your family must be. You should consider this promotion a well-deserved reward for your many years of untiring efforts in behalf of your company.

I rejoice with you in your good fortune. My very best wishes to you as you begin your new duties.

Cordially yours,

Y. S. Tsiang

註 ①hearty 衷心的 ②chief accountant 會計主任
③chemical 化學的 ④consider 認爲
⑤well-deserved 很値得的 ⑥reward 獎勵
⑦untiring 不倦的 ⑧effort 努力
⑨in behalf of 代表 ⑩rejoice 歡欣
⑪fortune 幸運 ⑫duties 責任，工作
⑬cordially 熱忱的

(五) 弔慰總經理逝世

Dear Sirs:

We are very sorry to receive the sad news of the passing of your General Manager Mr. Harry Schmid.

We fully realize how much the loss of Mr. Schmid will be felt by all of you and would like to extend our deepest sympathy on this sorrowful occasion.

Sincerely yours,

註 ①passing 逝世 ②general manager 總經理
③realize 明瞭 ④loss 損失
⑤felt 感覺到 ⑥extend 致送
⑦sympathy 同情 ⑧sorrowful 悲痛的
⑨occasion 事件

(六) 弔慰經理病故

Dear Mr. Chu:

I was greatly shocked yesterday to learn of your Manager's

death. I have known Mr. Sun almost from the moment he joined your organization ten years ago, and our close and friendly association was a source of enjoyment to me.

I know he was loved by his colleagues and he always had a friendly word for everyone. His cheerful disposition was an inspiration to all of us. He will be greatly missed by his many friends here.

Our thoughts are with you and we extend to you our deepest sympathy.

Sincerely yours,
S. T. Chao
Manager
Acme Company

註 ①shocked 震驚 ②death 死亡
③moment 時間 ④joined 加入
⑤organization 機構 ⑥close 親密
⑦friendly 友誼的 ⑧association 關係
⑨source 來源 ⑩enjoyment 愉快
⑪colleagues 同事 ⑫cheerful 歡欣
⑬disposition 性情 ⑭inspiration 啓示
⑮missed 懷念 ⑯thoughts 想念
⑰extend 致予 ⑱sympathy 同情

(七) 謝弔唁

Dear Sirs:

The Board of Directors and the General Manager of the Far Eastern Trading Co. acknowledge with sincere thanks your expression of sympathy addresses to them upon the decease of their esteemed and beloved Chairman, Mr. C. S. Lee.

Sincerely yours,

註 ①Board of Directors 董事會 ②acknowledge 覆告
③expression 表達 ④sympathy 同情
⑤addresses 言詞 ⑥decease 逝世
⑦esteemed 受尊敬的 ⑧beloved 被愛的
⑨Chairman 主席

(八) 慰病並祝早日康復

Dear Mr. Wells:

When I visited your office to place an order today, I was sorry to learn that you were out because of illness. I called the Veterans General Hospital immediately and was told that you are resting comfortably.

This is good news and I hope you will soon be your perky self again. Don't come back too quickly as we want to make sure that you are quite fit with a long, healthy career ahead of you.

Wishing you a speedy recovery.

Sincerely yours,

K. Lee

註 ①place an order 定貨 ②illness 病
③Veterans General Hospital 榮民總醫院
④resting 休養 ⑤comfortably 安適的
⑥perky 活潑的 ⑦fit 健全
⑧healthy 健康的 ⑨career 前程

⑩ahead 在前面　　⑪speedy 迅速
⑫recovery 痊癒

(九) 慰問友病

Dear John:

Since I had lunch with you only yesterday, I was greatly surprised and shocked when Steve told me this morning that you were in the hospital. He then explained that with a few weeks of rest you will be fit as ever.

Please take it easy before you get back to your hectic work. Get well soon. You are very much in the thoughts of all of us here, and you have our best wishes for a speedy recovery.

Sincerely yours,

Joseph Yu

註 ①lunch 午餐　　②surprised 驚異
③shocked 震驚　　④hospital 醫院
⑤explained 說明　　⑥rest 休養
⑦fit 健全　　⑧as ever 如舊
⑨take it easy 安心地　　⑩hectic 繁忙
⑪thoughts 想念　　⑫speedy 迅速
⑬recovery 康復

(十) 謝招待晚餐

Dear Mr. Gibson:

May I take this opportunity to thank you most sincerely for the very pleasant dinner which you so kindly gave me on the evening of March 10. It was one of the most enjoyable occasions during my stay in your city.

As you know it was my first visit to your country and I enjoyed very much the pleasant conversation I had with you and your colleagues.

Please give my best regards to Mrs. Gibson.

Sincerely yours,

D. C. Ho

註 ①opportunity 機會　②pleasant 愉快的
③dinner 晚餐　④enjoyable 歡欣的
⑤occasions 事件　⑥visit 訪問，觀光
⑦stay 居住　⑧conversation 談話
⑨colleagues 同事　⑩best regards 祝福

（十一）為女售貨員服務不週道歉

Dear Mr. King:

Thank you for your letter of May 5 calling our attention to the bad service of our sales-girl. Please accept our apology for the embarrassment caused you.

We want our service to be satisfactory to all our customers. Your letter affords us an opportunity to take action for the prevention of any such similar incident, and I am referring your letter to our Personnel Manager for his handling of the individual.

We hope this unhappy case will not deny the opportunity of continuing to serve you.

Yours truly,

註 ①calling 提醒　②attention 注意
③sales-girl 女售貨員　④apology 謝罪，歉疚
⑤embarrassment 煩惱　⑥caused 引起
⑦satisfactory 滿意　⑧affords 給予
⑨opportunity 機會　⑩take action 採取行動
⑪prevention 防止　⑫similar 相似的
⑬incident 偶發事件　⑭referring 交付予
⑮Personnel Manager 人事經理　⑯handling 處理
⑰individual 本人，個人　⑱unhappy 不愉快的
⑲deny 否定

（十二）預定旅館

Dear Sirs:

Please reserve for me a moderate bedroom and bath, preferably a southern exposure, beginning Sunday, April 5. I shall arrive early in the morning of the 5th and plan to leave in the afternoon of April 28.

Please confirm the reservation and also let me know the rate of the room.

Sincerely yours,
Nicholas Chow

註 ①moderate 普通的　②bedroom 臥室
③bath 浴室　④preferably 比較喜歡的
⑤exposure 面向　⑥plan 計畫
⑦leave 離開　⑧confirm 確定
⑨reservation 預定　⑩rate 房租

(十三) 覆告旅館已定妥

Dear Mr. Chow:

Thank you for your recent request. We are pleased to reserve for you an outside room with bath for April 5 to 28 inclusive. The rate will be US$ 45 a day.

We look forward to greeting you on your arrival.

Sincerely yours,

註 ①recent 最近 ②request 要求
③reserve 預定 ④inclusive 包括在內
⑤rate 房租 ⑥look forward to 盼望
⑦greeting 歡迎 ⑧arrival 抵達

(十四) 介紹友人請予協助

Dear Bill:

For a long time I have wished you to meet a very good friend of mine, Jack Wang, General Manager of the Taiwan Plastic Company.

Mr. Wang is visiting the United States to gather information about the effectiveness of his company's products. I am sure this is an area in which you can be of invaluable assistance to him.

As I believe that you both will not only be able to help each other but will also enjoy each other's company, I am writing this letter of introduction. I will, of course, appreciate any favor you can give him.

Sincerely yours,
Alex Cheng

註 ①General Manager 總經理 ②plastic 塑膠
③gather 搜集 ④information 資料，資訊
⑤effectiveness 效能 ⑥products 產品
⑦area 地區 ⑧invaluable 有價值的，寶貴的
⑨assistance 協助 ⑩each other 互相
⑪enjoy 享有 ⑫company 伴侶
⑬introduction 介紹 ⑭of course 當然
⑮appreciate 感謝 ⑯favor 恩惠，善意

（十五）因遷居退出中英文經協會

Dear Mr. Chang:

It is with sincere regret that I resign my membership in the Sino-British Cultural and Economic Association.

My company has move to Thailand and I am relocating my residence in Bangkok.

Please express my appreciation to the Board of Directors for the many courtesies they have shown to me and my family. I am deeply sorry to have to leave the Association and all the members with whom I have had the most rewarding relationship.

Sincerely yous,
William E. Ellen

註 ①regret 遺憾 ②resign 退出，辭職
③membership 會員
④Sino-British Cultural and Economic Association 中英文化經濟協會
⑤Thailand 泰國 ⑥relocating 移居
⑦Bangkok 曼谷 ⑧express 表示

⑨appreciation 感謝 ⑩Board of Directors 董事會

⑪courtesies 禮遇 ⑫family 家屬

⑬deeply sorry 十分抱歉 ⑭rewarding 有益的

⑮relationship 關係

第五章　信用狀

(LETTER OF CREDIT)

第一節　什麼是信用狀

在國際貿易上，除了商業書信之外，最重要而最常用的商業文件，便是信用狀了。

信用狀(Letter of Credit)也稱信用證；縮寫是L/C。定義如下：

“信用狀是一種用在國外貿易上的文件，通常由購貨人向所在地的銀行申請開發。內中說明購貨人同意付給售貨人或受益人 (Beneficiary) 一筆款項，但必須符合所規定的各種條件。信用狀由銀行根據顧客的請求開發，使售貨人增強了他對購貨人的信心。”

從上面的定義裏，我們可以知道下面的事實：

(1) 信用狀是由購貨人請求銀行開發的。

(2) 開發信用狀的銀行負有保證購貨人履行付款義務的責任。

(3) 購貨人規定在信用狀裏的條件，諸如貨物的數量、品質等等，必須完全符合後，出口方面的銀行才能將貨款付給售貨人。有了這種保障，可使交易順利達成，雙方滿意。

(4) 開發信用狀的銀行通知國外聯行依照條件付款，售貨人只須條件符合，就可以取得貨款，不必擔心貨款拖欠或拒付問題。

由於國際貿易間的買賣雙方，往往過去無往來，倘使沒有銀行居中服務，難以互相信賴。所以信用狀除買賣雙方各自負責而外，銀行也聯帶負了大部份的風險。

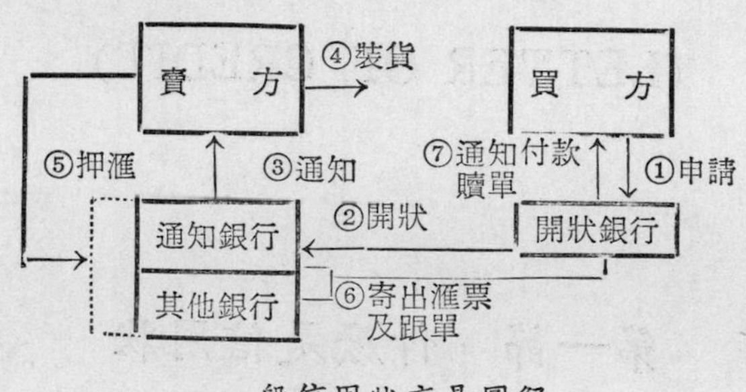

一般信用狀交易圖解

第二節　信用狀的種類

信用狀的種類很多，列舉如下：

(1)　商業信用狀。(Commercial letter of credit)
(2)　銀行信用狀。(Banker's letter of credit)
(3)　押滙信用狀。(Documentary letter of credit)
(4)　巡廻信用狀。(Circular letter of credit)
(5)　確認信用狀。(Confirmed letter of credit)
(6)　未確認信用狀。(Unconfirmed letter of credit)
(7)　可撤銷信用狀。(Revocable letter of credit)
(8)　不可撤銷信用狀。(Irrevocable letter of credit)
(9)　見票即付信用狀。(Sight letter of credit)
(10)　定期或承兌信用狀。(Time or Accepted letter of credit)

(11) 普通信用狀。(General letter of credit)

(12) 特殊信用狀（有限制條款）。(Special letter of credit)

(13) 可轉讓信用狀。(Transferable letter of credit)

(14) 不可轉讓信用狀。(Non-transferable letter of credit)

一般商業所用的信用狀， 都是“不可撤銷的確認商業信用狀” (Irrevocable confirmed commercial letter of credit, 簡稱 Irrevocable letter of credit.)

第三節 信用狀的內容

簡單地說，信用狀也就是銀行給售貨人的一封信。內容包括下列各項:

(1) 開發信用狀的銀行名稱。

(2) 開發信用狀的日期。

(3) 信用狀號碼及類別。

(4) 售貨人或受益人的名稱及地址。

(5) 申請開發信用狀人（購貨人）的名稱及地址。

(6) 貨價、購貨條件及付款條件。

(7) 所需隨同滙票(Draft)所附的各項文件（卽跟單）。

(8) 貨物名稱、數量及種類。

(9) 裝運條件；可否分批裝船，有無指定船公司等。

(10) 信用狀的限期。

(11) 滙票條款。

(12) 付款指示。

(13) 其他事項。

第四節 申請進口信用狀的手續

現在讓我們來詳細說明開發信用狀的手續。假定臺灣有一家公司要向外國一家公司購買紡織機器零件；那麼，首先要由購貨人向臺灣的銀行申請，然後由銀行用它的名義開出信用狀，透過它在外國的分行或聯號，發出通知書給外國的售貨人。

下面的例子，便是臺灣紡織公司向臺灣銀行申請開發信用狀的信和申請書的內容：

（甲）臺灣紡織公司給臺灣銀行的信

Bank of Taiwan　　　　April 6, 1989
Foreign Department
Taipei

Dear Sirs:

We are enclosing one copy of Application for Letter of Credit and shall appreciate your arranging to establish an L/C in Jardine's favor at your early convenience. Please note that all the details relating thereto are given in the Application.

Faithfully yours,

Encl. a/s　　　　Taiwan Textile Company, Ltd.

（乙）（附件）開發信用狀申請書（銀行印有現成格式）

APPLICATION FOR LETTER OF CREDIT

To: BANK OF TAIWAN　　　　Date: April 6, 1989
HEAD OFFICE FOREIGN DEPT.　　　　L/C No. 113/EU715
TAIPEI　　　　Amount £2, 300-0-0

Dear Sirs:

We request you to open Irrevocable Letter of Credit by airmail through your correspondent available by draft(s)drawn without recourse on your correspondent by Jardine Textile Machinery Works, Ltd., 4 Pall Mall, London, W. C. 1
for account of Taiwan Textile Company, Ltd.
at sight for any sum or sums not exceeding a total of Pounds Sterling Two Thousand and Three Hundred only
accompanied by the following documents:

1. Signed Commercial Invoice in quadruplicate indicating Import Licence No. MP-503
2. Insurance Policy (or Certificate) covering Marine Risks
3. Full set of Ocean Bills of Lading made out to order of BANK OF TAIWAN notifying ourselves
4. Others:

   Survey Report in duplicate

evidencing shipment of Spinning Machine Parts from Liverpool to Keelung (CIF) shipment to be made not later than December 31, 1981.
Partial shipments not allowed.
Bills of Lading indicating "On Deck" shipment not acceptable.
Insurance to be effected by shipper.
Freight to be prepaid by shipper.
Duplicate set of above documents to be sent by beneficiary to accountee directly.
Beneficiary's certificate to this effect required.
This credit is to remain in force until January 15, 1990.
SPECIAL INSTRUCTIONS:

L/C to be issued as marked "X"

☐	by	CABLE
X		AIRMAIL

【譯 文】

我們請求貴行代臺灣紡織公司用航空信經由貴聯行開發信用狀給倫敦樸爾麻爾四號渣甸紡織機器公司，以無追索權滙票方式見票即付任何金額不超過二千三百鎊，但須附有下列證件：

1. 已簽字的商業發票一式四份，註明進口證第MP-503號。
2. 保有海險的保險單（或證）。
3. 全套海運提貨單，抬頭人爲臺灣銀行轉知本公司。
4. 其他：公正人檢驗報告一式二份。

證明所運之紡績零件自利物浦至基隆 (CIF) 不得遲於一九八九年十二月三十一日。

不准分運。

提貨單倘註明貨裝在「甲板上」，不能接受。

保險由運貨人負責。

運費由運貨人預付。

上開證件由受益人以副本直接寄交購貨人。

有關此事的受益人證明書不可缺少。

本信用證有效期至一九九〇年一月十五日爲止。

特別指明：

信用狀照註明"X"的方格發出

☐	用	電 報
X		航空信

下面是印在申請書裏的條件，應由申請人簽字遵守：

1. In consideration of your granting above request we hereby bind ourselves duly to accept upon presentation and pay at your offices at maturity the drafts drawn under this Letter of Credit if the drafts and/or accompanying documents appear in the discretion of yourselves or your agents to be correct on their face.

2. And we agree to duly accept and pay such drafts even if such drafts and/or documents should in fact prove to be incorrect, forged or otherwise defective, in which case no responsibility shall rest with you and your agent.

3. And we further agree that you or your agents are not responsible for any errors or delays in transmission or interpretation of said Letter of Credit or for the loss or late or non-arrival of part or of all the aforesaid documents, or the quality, quantity or value of the merchandise represented by same, or for any loss or damage which may happen to said merchandise, whether during its transit by sea or land, or after its arrival or by reason of the non-insurance or insufficient insurance thereof or by whatever cause or for the stoppage, or detention thereof by the shipper or any party whomsoever, engaging ourselves duly to accept and pay such drafts in all like instances.

4. And we further agree that the title to all property which shall be purchased and/or shipped under this Letter of Credit the documents relating thereto and the whole of the proceeds thereof, shall be and remain in you until the payment of the drafts or of all sums that may be due on said drafts or otherwise and until the payment of any and all other indebtedness and liability, now existing or now

or hereafter created or incurred by us to you due or not due, it being understood that the said documents and the merchandise represented thereby and all our other property, including securities and deposit balances which may now or hereafter be in your or your branches' possession or otherwise subject to your control shall be deemed to be collateral security for the payment of the said drafts.

5. And we hereby authorise you to dispose of the afore-mentioned property by public or private sale at your discretion without notice to us whenever we shall fail to accept or pay the said drafts on due dates or whenever in your discretion it is deemed necessary for the protection of yourselves and after deducting all your expenses to reimburse yourselves out of the proceeds.

6. And in case of extension or renewal of this Letter of Credit or modification of any kind in its terms, we agree to be bound for the full term of such extension or renewal, and notwithstanding any such modification.

7. And in case this request is signed by two or more, all promises or agreements made hereunder shall be joint and several. We herewith bind ourselves to settle exchange on drafts drawn under this Letter of Credit with your goodselves.

Yours faithfully,
Taiwan Textile Company
C. C. Hsu, Manager

【譯　文】

(1) 關於本信用狀下的滙票及其附屬單據等，如經貴行或貴代理行認爲在表面上尙屬無誤，敝處於滙票提示時即承兌並依期照付。

(2) 上項滙票單據等，縱然在事後證實爲非真實或僞造或有其他缺點，概與貴行及貴代理行無涉。

(3) 本信用狀的傳遞錯誤或遲延或解釋上的錯誤，及關於上面單據所載貨物或貨物的品質或數量或價值等，有全部或一部份滅失或遲遞，或因沒有抵達交貨地，以及貨物無論因在海面或陸上運輸中，或運抵後，或未經保險，或保額不足，或因承辦商或任何第三者的阻滯或扣留，及其他因素等情，以致喪失或損害時，與貴行或貴代理行無涉。在以上任何情形之下，該滙票仍應由敝處兌付。

(4) 與上述滙票及與滙票有關的應付各款，以及敝處對貴行，不論現已發生，或日後發生，已經到期，或尙未到期的其他債務，在未清償以前，貴行得就本信用狀項下所購運的貨物單據，及賣得價金，視同如自己所有，並應連同敝處所有其他財產，包括存在貴行及分支機構，或貴行所管轄範圍內的保證金存款餘額等，均任憑貴行移作上述滙票的共同擔保，以備作淸償票據之用。

(5) 如上述滙票到期，而敝處不能照兌時，或貴行因保障本身權益，認爲必要時，貴行可以不經通知，有權決定，將上述財產（包括貨物在內），以公開或其他方式，自由變賣，在賣得價金內，扣除費用後，抵償貴行借墊各款，無須另行通知敝處。

(6) 本信用狀如經展期或重開及修改任何條件，敝處對於以上各款，絕對遵守，不因展期、重開、或條件的修改而發生異議。

(7) 本申請書的簽署人，如果是二人或二人以上時，對於申請書所列各項條款，自當共同連帶及個別負擔全部責任，並負責向貴行辦理一切結滙手續。

臺灣紡織公司

經理 徐 承 照

另有保證書，由保證人簽字：

GUARANTEE

（保 證 書）

We hereby guarantee jointly and severally the fulfilment of the promises and agreements, contained herein, including extensions, renewals and modifications, and in the event of default promise to make good and pay on demand any loss or damage suffered by the BANK OF TAIWAN, waiving hereby expressly any defence that may be interposed to any claim or action thereon or hereon; especially also as to the order in which the BANK OF TAIWAN shall choose to reimburse itself.

Nancy Chow of Fair Co.
(Guarantor)

【譯　文】

本公司願和申請人共同連帶，並各個單獨負責保證履行上開申請書所列各項承諾及協議行爲，包括展期、重開、及修改事項等行爲在內。如有違約情事，願立即賠償貴行所受之一切損失，並對貴行求償損失之任何要求，或訴訟行爲，聲明，均放棄先訴抗辯權。特此保證。

飛亞公司　周　繼　慧
（保證人）

臺灣銀行接到臺灣紡織公司的信及申請書以後，倘無問題，就透過倫敦聯行開發下面的信用狀給售貨人 Jardine 紡織機器公司：

（丙）信用狀

BANK OF TAIWAN

Advised through

National Provincial Bank, Ltd.
London W. C. 1
England

To:

Jardine Textile Machinery Works, Ltd.　　April 7, 1989

4 Pall Mall
London W. C. 1
England

Irrevocable Credit
No. 113/EU715
Amount £2, 300-0-0

We hereby authorize you to draw on National Provincial Bank, London, W. C. 1, England for account of Taiwan Textile Company, Ltd., 43 Nanking W. Road, Taipei, Taiwan up to an aggregate amount of Pounds Sterling Two Thousand Three Hundred only available by your drafts at sight for full invoice value accompanied by the following documents:

Signed Commercial Invoice in quadruplicate, indicating Import License No. MP-503.

Marine Insurance Policy or Certificate in duplicate, endorsed in blank, in currency of drafts for 110% of the invoice value including institute cargo clause, institute war clause.

Supplier's Certificate filled in and signed (form attached)

Your certificate that you have forwarded one orginal Bill of Lading, Packing List, Consular Invoice and Commercial Invoice direct to the accountee

Survey Report in duplicate

relating to

SPINNING MACHINE PARTS

Shipment from Liverpool to Keelung

Partial shipments prohibited. Transhipment permitted.

Drafts must be presented for negotiation not later than October 15, 1989.

Drafts must bear the above credit number and the amount negotiated must be endorsed on the reverse hereof by the negotiating bank.

We hereby agree with the drawers, endorsers and bona fide holders of the drafts drawn under and in compliance with the terms of this credit, that such drafts will be duly honored upon presentation to the drawee bank.

Negotiating banks must forward the drafts to the drawee bank for reimbursement and airmail the documents to us.

All banking charges outside Taiwan are for your account.

Your faithfully,

For BANK OF TAIWAN

【譯 文】

茲授權貴公司向倫敦國家省銀行開出臺灣紡織公司抬頭的即期滙票，金額總數爲二千三百鎊的全部發票價值，連同下列單據：

已簽字的商業發票，一式四份。

海運保險單一式兩份，空白背書，照發票價的滙票百分之一百十投保，包括貨物條款及戰爭條款。

供貨人的證明，在附表上塡寫及簽字。

貴公司證明已將原提單，包裝單，領事發票及商業發票各一份直接寄交購貨人。

公正人檢驗報告一式二份。

有關

紡紗機零件

自利物浦運往基隆。

不准分運。可准轉船。

滙票應在一九八九年十月十五日以前洽兌。

滙票上應註明本信用狀號數。

洽兌款數應由承兌銀行在反面背書。

本銀行與遵守本信用狀條件的發票人、背書人、及持有人，一致同意此項滙票，一經提交，應由付款銀行照付。

各承兌銀行必須將滙票寄交付款銀行，以取回墊款，並將單據寄交本行。在臺灣以外的一切費用，由貴公司負擔。

臺灣銀行 啓

註 一般說來，貨物的投保，應該在金額上是加百分之十的，所以信用狀裏說要照貨價保 110%。這附加的投保額是保障費用上的損失，因爲萬一貨物全部損失了，雖然可以得回貨値，但我們却要損失運費、利息，銀行費用等，因此投保時必須加百分之十，作爲保障。

同時臺灣紡織公司也用下面的信通知 Jardine 紡織機器公司，說明信用狀已由臺灣銀行開出：

April 9, 1989

Jardine Textile Machinery Works, Ltd.

4 Pall Mall

London W. C. 1

England

Re: Spinning Machine Parts

Dear Sirs:

Your letter of March 4, 1981 confirming the order we placed with you for the subject parts has been received. We are pleased to inform you that our applications for both import license and letter of credit have been approved.

On April 7 an irrevocable letter of credit was opened in your favor for an amount of £2, 300-0-0 to cover the CIF value of this order by Bank of Taiwan through the National Provincial Bank, London. For your information, the numbers of the import license and the L/C are MP-503 and 113/EU715 respectively. It will be appreciated if you will arrange to ship the total quantity three months after receipt of the L/C.

Enclosed for your reference is one copy of the L/C we have opened for this order.

Faithfully yours,

Taiwan Textile Company, Ltd.

Encl. a/s

註 ①confirm 確定　②application 申請

③import licence 進口證　④approve 批准

⑤respectively 各別地

倘使原信用狀內容以後有修改的必要，應由臺灣紡織機器公司再函請臺灣銀行申請修正。下面便是請求修改貨價的出口商來信，以及進口商的申請函和申請書：

(甲) 出口商要求更改貨價函

Taiwan Textile Company, Ltd.　May 4, 1989

3rd Floor, 43 Nanking W. Rd.

Taipei, Taiwan

Re: L/C No. 113/EU715

for Spinning Machine Parts

Dear Sirs:

We acknowledge with thanks the receipt of your letter of April 9, 1981 and wish to inform you that we have already received the captioned letter of credit.

Much to our regret, the wage rates of workers at our plants have increased by 10% across the board effective from May 1, 1981. The wage boost was finalized at a meeting between our management and the labor union on April 20th, in order to head off a strike scheduled on April 30th. In view of the fact that such increase is beyond our control, we are now compelled to request you to increase the amount of the subject L/C by £230-0-0. Your compliance with our request will be greatly appreciated.

Faithfully yours,
Jardine Textile Machinery Works, Ltd.

註 ①much to our regret 我們十分遺憾

②wage rate 工資率 ③plant 工廠

④across the board 全部 ⑤effective 生效

⑥boost 增加 ⑦finalize 最後決定

⑧management and labor union 勞資雙方

⑨head off 阻止 ⑩strike 罷工

⑪scheduled 預定 ⑫beyond our control 非我們所能控制

⑬are compelled 被迫 ⑭compliance 照辦

(乙) 進口商申請更改信用狀函

Bank of Taiwan May 12, 1989
Head Office Foreign Department
Taipei

Re: L/C No. 113/EU715

Dear Sirs:

We have just received a letter from the beneficiary of the subject L/C, Jardine Textile Machinery Works, Ltd., requesting us to increase the amount thereof by £230-0-0 owing to a recent increase of wage rates at their plants.

Enclosed please find one copy of Application for Amendment of Credit for your processing. It will be appreciated if you will amend the credit and advise your correspondent in London accordingly.

Faithfully yours,

Taiwan Textile Company, Ltd.

Encl.: a/s

註 ①beneficiary 受益人 ②amendment 更改

③processing 辦理 ④correspondent 聯行

(丙)(附件)更改信用狀申請書

BANK OF TAIWAN

Inward Bills Division

Head Office–Foreign Department

TAIPEI, TAIWAN

APPLICATION FOR AMENDMENT OF CREDIT

Amendment No. ________________ Date: May 12, 1989

Credit No. 113/EU715

Import License No.: Original MP-503; Amended________________

Amount: £2,300-0-0

Shipper: Jardine Textile Machinery Works, Ltd.

Advising Bank: National Provincial Bank, London

Please amend following by AIRMAIL

Alteration to be made:	ITEM Originals stipulated in the credit:
£2,530-0-0	AMOUNT £2,300-0-0
....	SHIPPER'S NAME AND ADDRESS ....
....	....
....	BUYER'S NAME AND ADDRESS...........
....	....
....	COMMODITY AND QUANTITY..............

.... PRICE TERM..

.... SHIPPING PORT ....

.... PORT OF ENTRY.....................................

.... SHIPMENT TO BE MADE BY................

.... DATE OF SHIPMENT..............................

.... DATE OF EXPIRY ....

.... OTHERS

Signature of Applicant

Taiwan Textile Company, Ltd.

註 ①alteration 更改 ②stipulate 規定

③date of expiry 到期日

（丁）臺灣紡織公司函告 Jardine 紡織機器公司

Jardine Textile Machinery Works, Ltd. May 14. 1989

4 pall Mall

London, W. C. 1

England

Re: L/C No. 113/EU715

Dear Sirs:

Your letter of May 4, 1981 requesting us to increase the amount of the captioned L/C by 10% has been received.

To comply with your request, we submitted our Application for Amendment of Credit to the opening bank on May 12th, and trust you will receive from the advising bank the amendment soon.

As we have informed you in our previous letters, we are in urgent need of these parts for replacements. Therefore, you are requested to ship them as originally scheduled.

Faithfully yours,

Taiwan Textile Co., Ltd.

註 ①request 請求　②comply with 遵照

③submit 致送　④previous 以前的

⑤in urgent need 急需　⑥replacement 補充

⑦as originally scheduled 照原定日期

如果一切順利，在貨物由倫敦運來臺灣的時候， Jardine 紡織機器公司就可以憑各項文件單據，向臺灣銀行的倫敦聯行取款。貨物到了臺灣以後，臺灣銀行隨卽通知臺灣紡織公司付款贖單。然後再憑單去到海關辦理提貨手續。全部過程，到此結束。

第五節　出口信用狀

以上說明了申請進口信用狀的手續。現在再來談出口信用狀。假定西德有家公司向臺灣買了六十噸的罐頭蘆筍。按照手續，西德的進口商必須向西德的銀行申請開發信用狀。但爲了爭取時間，他先請銀行以電報通知臺北聯行，再補發信用狀。擧例如下：

（甲）西德進口商申請以電報開發信用狀

APPLICATION FOR IRREVOCABLE CREDIT

The Manager,　September 10, 1989

Hamburg Commercial Bank

HAMBURG

Dear Sirs:

Please establish by CABLE an Irrevocable Credit in the following terms through your office or correspondents at Taipei, Taiwan, in favour of General Trading Company, Ltd., P. O. Box 123, Taipei, Taiwan. Draft (s) to be drawn on Universal Importers, Inc., P. O. Box 320, Hamburg, West Germany.

Up to an amount not exceeding US$ 3, 800. 00 (say U. S. Dollars Three Thousand and Eight Hundred Only.) at sight for full invoice value of the goods specified below and accompanied by the following documents which are to be surrendered against payment of the draft(s):

(1) Signed invoice (s) in quadruplicate.

(2) Complete set of clean "Shipped on Board" Ocean bills of lading marked freight prepaid made out "to order" and endorsed in blank.

(3) Certificate of Origin in triplicate.

evidencing shipment from Keelung to Hamburg not later than September 30, 1981 of the goods specified below.

Partial shipments are not allowed. Transhipment is prohibited.

Draft (s) drawn under this Credit must be presented for negotiation in Taipei on or before October 30, 1981.

PARTICULARS OF GOODS:

Canned Asparagus, Al Grade

Tips and Cuts	20 metric tons
Center Cuts	20 metric tons
End cuts	20 metric tons

Shipping Marks:-

OTHER TERMS:

(1) Insurance to be covered by drawees.

(2) Shipper to advise buyer by cable after shipment is effected, the name of vessel, date of sailing, total quantity and the number of packages and a copy of cable evidencing same should be attached to other documents.

Signature of Applicant: Universal Importers, Inc.

註 ①establish by cable 用電報開發 ②at sight 見票即付
③specify 規定 ④accompany 隨附
⑤document 文件（跟單） ⑥surrender 繳出
⑦quadruplicate 一式四份 ⑧clean 未塗改或批註的
⑨bill of lading 提貨單 ⑩endorse 背書
⑪blank 空白 ⑫certificate of origin 產地證明書
⑬evidencing 證明 ⑭partial shipment 分批交運
⑮transhipment 轉船 ⑯prohibit 禁止
⑰negotiate 承兌 ⑱drawee 收款人

西德漢堡商業銀行接到上面的申請書後，除一面先發電報外，並開發下面的信用狀，透過它的臺北聯行臺灣銀行通知臺北的出口商。

（乙）信用狀

HAMBURG COMMERCIAL BANK
P. O. Box 573
Hamburg, West Germany

GENERAL TRADING COMPANY, LTD.
P. O. Box 123
Taipei, Taiwan

September 16, 1989
VIA CABLE THROUGH
BANK OF TAIWAN
TAIPEI, TAIWAN

Dear Sirs:

IRREVOCABLE LETTER OF CREDIT NO. 51283

We hereby establish our irrevocable letter of credit in your favour for Account of UNIVERSAL IMPORTERS, INC., HAMBURG

Up to an aggregate amount of US$3,800.00 (United States Dollars Three Thousand and Eight Hundred Only)

Available by your drafts at sight drawn without recourse on the accountee for full invoice value accompanied by the following documents:

1. Signed Invoice in quadruplicate
2. Packing List in quadruplicate
3. Certificate of Origin in triplicate
4. Full set of clean "On Board" Ocean Bills of Lading made out to Order of Hamburg Commerical Bank, Hamburg notify Buyers marked "Freight Prepaid"

Evidencing shipment of:

Sixty (60) metric tons of Canned Asparagus, Al Grade

C. & F. HAMBURG

Shipment from Taiwan to Hamburg not later than September 30, 1981.

Partial shipments are not permitted. Transhipment is prohibited.

All bank charges outside West Germany are for account of beneficiaries.

Drafts drawn under this Credit are to be negotiated at sight basis, discount charges are for drawees' account.

Negotiating banks to forward all documents direct to us by airmail.

This credit expires on October 30, 1981 for negotiation in Taiwan.

Yours faithfully,

For HAMBURG COMMERCIAL BANK

註 ①aggregate amount 總數　　②without recourse 無追索權

（丙）臺灣銀行根據電報通知臺北的出口商

BANK OF TAIWAN

HEAD OFFICE

FOREIGN DEPARTMENT

Our Ref. No. 70/5213　　Taipei, Taiwan September 17, 1989

TO:

GENERAL TRADING COMPANY, LTD.
P. O. Box 123
Taipei

Original credit issued by

(Their Ref. No. 3513　　Dated

Advised by: (Ref. No.　　Dated

Hamburg Commercial Bank

Hamburg, West Germany

Amount: US$3,800.00
Shipping date:
Expiry date:
Covering:

Dear Sirs:

We wish to inform you that we have received a cable from our correspondent (IT 12179) dated September 15th which decoded reads as follows:

ORDER UNIVERSAL IMPORTERS HAMBURG AIRMAILING CREDIT 51283 USDOLLARS 3800 FAVOR GENERAL TRADING COMPANY LTD PO BOX 123 TAIPEI COVERING 60 METRIC TONS CANNED ASPARAGUS AI GRADE

Please note that this is merely an advice on our part and does not constitute a confirmation of this credit. This bank is unable to accept any responsibility for errors in transmission or translation of this cable or for any amendments that may be necessary upon receipt of the mail advice from the bank issuing this credit.

Kindly acknowledge receipt by returning the attached form duly signed, and oblige.

Yours faithfully,

For BANK OF TAIWAN

註 ①covering 包括 ②decode 由密碼譯出
③canned asparagus 罐頭蘆筍 ④merely 僅僅地
⑤constitute 構成 ⑥transmission 傳遞
⑦amendment 更改 ⑧attached form 隨附表格

上面這一張信用狀，和我們前面所提到的入口手續所用的信用狀差不多是一樣的。分別的地方在於入口信用狀是我們的入口商委托銀行開發給出口商；出口信用狀則由外國銀行代入口商開發給我們的出口商。此外一是用航空信開的信用狀，另一先用電報開的信用狀。至於信用狀的形式，並不是每家銀行都是一樣，但內容大致相同。（參閱本章所附信用狀舉例）

出口商收到信用狀以後， 必須仔細核閱。 因爲信用狀內， 如果有少許錯誤而我們不立即要求西德的買主通知開信用狀的銀行更正的話， 等到我們將貨物付船後拿單據到銀行取款， 便會發生很大的麻煩。縱然是一字之微，銀行也會拒絕通融。但是信用狀包含的項目相當多，如果沒有一個好的辦法去核對，很容易會有遺漏的地方。最理想的辦法是準備一張核對單 (Check List)，以便逐項一一檢查。核對單可分列下面各項:

(1) 信用狀受益人的名字(Beneficiary's name)。

(2) 貨物種類(Description of goods)

(3) 數量和單價(Quantity & unit price)。

(4) 信用狀金額(Amount)。

(5) 付船日期(Shipping date)。

(6) 付船港口(Shipping port)。

(7) 目的地(Destination)。

(8) 分期付船是否同意(Partial shipment allowed or not)。

(9) 轉船是否同意(Transhipment allowed or not)。

(10) 所需單據文件(Documents required)。

(11) 有效終止日期(Expired date)。

(12) 其他特定條款(Other special terms)。

(13) 所開來的是何種信用狀(Which kind of L/C)。

除此而外，必須再拿出西德買方寄來的定貨單，和我們寄給他們的銷售合約，一起來詳細覆核。

現在再來說明我們出口商應該怎樣去做核對工作。

(一)信用狀受益人的名字 收到西德入口商寄來的信用狀，受益人的名字當然是我們了。但是上面的名字，是不是和我們公司的名字完全一樣呢？例如我們公司的名字是 "General Trading Co."， 而信用狀上所列的却是 "General Trading Corp."。看來是差不多，但如果不通知西德買方更改，我們就無法向銀行取得貨款。

(二)貨物種類 例如售貨合約上列明的是女用雨傘，倘使信用狀上只寫"雨傘"，這樣便不符規定，非通知買方更改信用狀不可。

(三)數量和單價 這兩點也應和合約核對，最要緊的是檢查一下價格條件是否與原約相符？ 例如保險費、 運費等應該由何方負擔等等。

（四）金額　核對金額時，我們必須將單價和數量核算。以及應否扣除佣金。

（五）裝船日期　在核對信用狀時，對於裝船日期應當特別注意。假使入口商規定我們的裝船日期使我們無法接受，便應該立即商請買方改期，並請通知銀行更改信用狀。

（六）裝船港口　除了注意裝船日期之外，還要注意信用狀上所規定的裝船港口。例如指定由基隆港裝船的，便不能在高雄港裝運。

（七）目的地　信用狀上規定的目的地，並不一定是入口商的所在地。例如倫敦一家進口商向臺北定購一千噸水泥，但目的地並非倫敦；可能是要運往法國的馬賽。所以核對時應特別小心。

（八）分期付船是否同意　有些入口商對於分期交貨是同意的。但是縱然在合約上訂明，如果沒有在信用狀上註明，我們就不能分期裝船。

（九）轉船是否同意　當然，出口商和入口商都希望貨物直接運到目的地，但有時的實際情形却非轉船不可。例如臺灣只有船去香港，而沒有船去澳門。可是香港和澳門之間却有船往來。這樣，澳門入口商向臺灣定購的貨物，一定要經香港轉船。我們核對信用狀，如果上面說明轉船不同意，我們便要向船公司調查一下，有沒有直接到達澳門的船。如果沒有，應立即通知進口商更改信用狀，准許轉船。否則便無法交貨。

（十）所須單據文件　進口商在開發信用狀時，最重要的便是這一項。他們在信用狀上面所規定必需的單據，諸如發票、保險單、滙票、提貨單等，是用來約束出口商怎樣交貨的。因此，我們應該根據合約詳細一一核對。如果不能同意，便立即通知買方更改。例如我們的賣價是C. & F.（貨價及運費），倘使對方在信用狀所列單據中包括保險單在內，便不合理了。再如他們在沒有同意付出公正人檢驗費用，而合約裏又沒有規定賣方要付檢驗費用，假使在信用狀裏，他們規定要附有公正人的檢驗證明書，我們也有理由拒絕的。

(十一)有效終止日期 這是規定信用狀的最後一天有效日期。換句話說，過了這一天，信用狀便無效了。出口商縱然在信用狀有效期以前將貨運出，如果到銀行洽領貨款在有效期以後，銀行也不會接受的。

(十二)其他條款 有時進口商在信用狀上加上其他特別條款，例如貨物不能交付某家輪船公司裝運，或裝船後立即用電報通知等等。對於這一類的特別條款能否接受，我們在核對信用狀時，必須特加注意。倘使不能同意，應迅速通知對方修改。

(十三)所開來的是何種信用狀 這一項也是十分重要的。我們所要的信用狀，應該是"不可撤銷的"保證付款的商業信用狀。倘使是可撤銷的，那就無法接受了。

在逐項核對信用狀清楚以後，如果沒有需要更改的地方，我們就應該準備履行合約，如期裝船，在信用狀有效日期以前，向銀行憑單據文件及滙票，收取貨款。

臺灣銀行付款以後，它又將全部單據及滙票寄給西德的聯行，向該行接洽收回所墊付的款項。

信用狀(例一)

BANK OF CHINA

Taipei, Taiwan

Republic of China

Irrevocable Letter of Credit No. 57103 Date: September 8, 1989

Amount: US$585.00

To: Mizobata & Co., Ltd. Advised through

750 Nishi Kaizuka
Osaka, Japan

BANK OF CHINA, Osaka, Japan

[x] This is the original credit opened by AIRMAIL

[ ] This is a confirmation of our credit opened by CABLE today

Dear Sirs:

We hereby authorize you to draw at sight on our Osaka Branch Office for account of Formosa Textile Company, Ltd., P. O. Box No. 23054, Taipei, Taiwan to the extent of U. S. Dollars Five Hundred Eighty-Five only advailable by your draft(s) accompanied by:

1. Commercial Invoice in quadruplicate (three copies of which must be forwarded by airmail with original set of documents) duly signed, Indicating Import License No. MP-903 and number of this Credit, and showing breakdown of cost, insurance premium and freight in the case of CIF or C&F shipments.
2. Consular involce (duplicate copy acceptable)
3. Full set of clean "On Board" ocean bills of lading, marked freight prepaid, made out to order of Bank of China, H. O. F. D. Taipei notify buyer.
4. Supplier's certificate as indicated on the reverse side of this Credit, which must be duly filled out(blank form unacceptable.)
5. Others: Certificate of Origin in quadruplicate.

Evidencing shipment of

1,000 kgs. each of Types Nos. 77 and 100-110 of "Jet Size" Sizing Stuff.

from Osaka, Japan to Keelung

Shipment to be made not later than October 20, 1981.

Partial shipments not allowed.   Transhipment not allowed.

Bills of Lading indicating "On Deck" shipments are not acceptable

Insurance to be effected by Buyer

Freight to be prepaid by shipper

Buyer's full address shown herein must be indicated on shipping documents.

Special Instructions:

Draft (s) must indicate: "Drawn under Bank to China, Head Office-Foreign Department, Taipei, Irrevocable Letter of Credit No. 57103 dated September 8, 1980"

The amount and date of each negotiation must be endorsed on the back hereof by the negotiating bank.

We hereby agree with the drawers, endorsers, and bona fide holders of the draft(s) drawn under and in compliance with the terms of this Credit that such draft (s) shall be honored on due presentation and delivery of documents as specified, if presented for negotiation on or before October 31, 1989.

Unless otherwise expressly stated, this credit is subject to the "Uniform Customs and Practice for Documentary Credits (1962 Revision), International Chamber of Commerce Brochure No. 222."

Faithfully yours,
For BANK OF CHINA
Head Office-Foreign Department

信 用 狀(例二)

THE CHARTERED BANK

INCORPORATED IN ENGLAND BY

ROYAL CHARTER 1853

TO CENTRAL TRUST OF CHINA
49, Wu Chang Street, Sec. I,
TAIPEI, TAIWAN.

Bangkok, 30th MARCH, 1989
Import licence No. 402576.
PRE/B. 430. 660. 901. 456. 4260/0147.
Irving Trust Co., New York.
T. C. D.: 31st August, 1989.
Letter of Commitment: 430-S-4541.

IRREVOCABLE LETTER OF CREDIT No. 61/2276-z
WITHOUT RECOURSE TO DRAWERS

Dear Sir (s):

You are hereby authorised to draw without recourse to you on THISETUONG, 2001, Thanh Rd. Bangkok (Nguyen-Thi-Du) domiciled with us, ...for a sum not exceeding US$4, 233.-C. &F. (say U. S. Dollars four thousand two hundred and thirty three only.) .... available by your drafts, drawn in duplicate, on them at sight accompanied by the following documents:

Complete set of not less than two clean Ocean Bills of Lading to order blank e n do rsed.

"Shipped on Board" Bills of Lading are essential and the statement "Freight paid" must appear thereon. The Bills of Lading must specify shipment as detailed below.

—Insurance covered in Bangkok

+Short form of Lading are not acceptable.

—Further instructions see reverse.

Signed Invoices in duplicate,

Visaed by Chamber of Commerce.

Certificate of Origin Taiwan, in duplicate, visaed by Chamber of Commerce evidencing shipment from TAIWAN

to BANGKOK

of the following merchandise: —35 M/T STEEL ANGLE BARS

in length of 6 meters

FOB......US$3,715.00

FREIGHT 518.00

C. & F. US$4,233.00

Condition of Shipment. Total shipments by U. S. vessel

Transhipment is permitted.

Shipment date. Shipment is to be effected not later than May 1, 1981

Partial/shipments are permitted.

Expiry Date. This Credit expires on 30th JUNE 1981 for negotiation in TAIWAN. Drafts should bear the following clause "drawn under The Chartered Bank BANGKOK Credit No. 61/2275-z; date 30th MARCH 1981." Purchasers are to note the amount of the drafts separately on the back hereof. We hereby engage with the drawers, endorsers and bona fide holders of bills drawn and presented in accordance with the terms of this credit that the bills shall be duly honoured on presentation.

—Shipments of a value inferior

to US$1,000.00 are not permitted except for the last shipment.

Yours faithfully,

____________Manager/No. 46

____________Accountant/No. 155

信 用 狀 (例三)

BANK OF AMERICA
NATIONAL TRUST AND SAVINGS ASSOCIATION

IRREVOCABLE LETTER OF CREDIT

San Francisco, California November 13, 1989

Kingstom Exporters ALL DRAFTS DRAWN MUST BE MARKED:
Kingstom DRAWN UNDER "B OF A" CREDIT NO. 29330

GENTLEMEN:

WE HEREBY ESTABLISH OUR IRREVOICABLE LETTER OF CREDIT IN YOUR FAVOR FOR ACCOUNT OF Royal Packers, Santa Clara, California UP TO THE AGGREGATE AMOUNT OF US$3,900.00(THREE THOUSAND NINE HUNDRED U. S. DOLLARS) AVAILAB LEBY YOUR DRAFTS DRAWN AT SIGHT AND ACCOMPANIED BY DOCUMENTS SPECIFIED

BE: ON COVERING______full______INVOICE VALID OF MERCHANDISE TO BE DESCRIBED IN INVOICE AS:

Twelve Long Tons Jamaican Pimientos, F. O. B. Kingston.

DOCUMENTS REQUIRED:

1. Commercial invoice in triplicate.
2. Special U. S. Customs invoice in duplicate.
3. Full set of clean on board ocean Bills of Lading, to order of shipper, blank endorsed, marked: Notify Royal Packers, Santa Clara".

Shipment latest December 15, 1980 from Kingston, Jamaica to San Francisco, Calif. Partial shipments are permitted. Transhipment is permitted. Insurance to be effected by the buyer.

The negotiating bank is hereby authorized to make advances to the beneficiary up to the aggregate amount of US$3,900.00 or the remaining unused balance of this Credit (whichever is less) against the beneficiary's receipt stating that the advances are to be used to pay for the purchase and shipment of the merchadise is covered by this Credit and the beneficiary's undertaking to deliver to the negotiating bank the documents stipulated in the Credit. The advances with interest are to be deducted from the proceeds of the drafts drawn under this Credit. We hereby undertake the payment of such advances with interest should they not be paid by the beneficiary prior to the expire action of this Credit.

THE AMOUNT AND DATE OF EACH NEGOTIATION MUST BE ENDORSED ON THE BACK THEREOF BY THE NEGOTIATING BANK. WE HEREBY AGREE WITH BONA FIDE HOLDERS THAT ALL DRAFTS DRAWN UNDER AND IN COMPLIANCE WITH THE TERMS OF THIS CREDIT SHALL MEET WITH DUE HONOR UPON PRESENTATION AND DELIVERY OF DOCUMENTS AS SPECIFIED TO THE DRAWEE, IF DRAWN AND NEGOTIATED ON OR BEFORE December 28, 1981.

PROVISIONS APPLICABLE TO THIS CREDIT "UNLESS OTHERWISE EXPRESSLY STATED, THIS CREDIT IS SUBJECT TO THE UNIFORM CUSTOMS AND PRACTICE FOR COMMERCIAL DOCUMENT CREDITS FIXED BY THE THIRTEENTH CONGRESS OF THE INTERNATIONAL CHAMBER OF COMMERCE."

Yours faithfully,

Bank of America
National Trust and Savings Association

BY____________________ BY____________

AUTHORIZED COUNTER SIGNATURE MANAGER-ASSISTANT
VICE PRESIDENT
ASSISTANT CASHIER

信 用 狀（例四）

NATIONAL BANK OF NORTH AMERICA

Nov. 21, 1989

IRREVOCABLE COMMERCIAL LETTER OF CREDIT

Carline Plastic Products Mfg. Co., Ltd. Advised through
First National City Bank

Gentlemen:

We hereby authorize you to draw on National Bank of North America, New York. N. Y. by order of A. J. Industries Inc. 140 Adams Blvd. Farmingdale. New York, 11835, and for account of same up to an aggregated amount of $15,584.36 U. S. Currency (Fifteen Thousand Five Hundred Eighty-Four and 36/100 U. S. Dollars). Available by your drafts at 60 days sight for 100% of the invoice value. Accompanied by:

Commercial Invoice in quadruplicate

Special Customs Invoice in duplicate

Full set clean on board ocean Bills of Lading, issued to the order of National Bank of North America, marked notify buyer.

EVIDENCING SHIPMENT OF: Vinyl Inflatable Toya, as per Pro Forma Invoice dated November 8, 1981.

From: Taiwan to New York, New York

Insurance coverd by buyer.

Partial shipments not permitted.

The negotiating bank must forward all documents to us in one airmail.

Drafts must be drawn and negotiated not later than Dec. 15, 1989.

All drafts must be marked "Drawn under the National Bank of North America Credit No. A 23056 and the amount endorsed on this L/C.

We hereby agree with the drawers, endorsers, and bona fide holders of all drafts drawn under and in compliance with the terms of this credit, that such drafts will be duly honoured upon presentation to the drawee.

Unless otherwise expressly stated, this Credit is subject to the Uniform Customs and Practice for Commercial Documentary Credits (1962 Revision) the International Chamber of Commerce Brochure No. 222.

Yours very truly,

NATIONAL BANK OF AMERICA

信用狀(例五)

IRREVOCABLE WITHOUT RECOURSE

LETTER OF CREDIT

THAI DEVELOPMENT BANK, LTD.

Head Office

276.278 Tajawongse Road, Bangkok

Cable Address
"Patanabank"
P. O. Box 75
Code Use:
Peterson
International Code

Whinlin Electric & Engineering Corp.,
66, Nanking Road, West,
Taipei, Taiwan

IRREVOCABLE LETTER OF CREDIT

NO. 1254/67. FOR US$ 3, 997. 40

Dear Sirs:

We hereby authorize you to draw without recourse on Messrs.

Lopburi Co., Ltd., Bangkok, to the extent of US$3,997.40 (Say U. S. Dollars Three Thousand Nine Hundred Ninety-seven and Cents Forty Only). for 100% invoice value to be accompanied by complete set of documents as specified:-

DRAFTS in duplicate drawn at sight bearing the clauses "DRAWN UNDER THE THAI DEVELOPMENT BANK LTD., BANGKOK L/C No. 1254/67, dated September 7, 1989, Document against payment." Signed Commercial Invoice in six-duplicate.

Marine Insurance Policy or Certificate in duplicate, endorsed in blank, for 110% of Invoice value covering:

Institute War Clauses.

Institute Cargo Clauses. (All risks)

Institute Strikes, Riots & Civil Commotions Clauses.

Institute Theft, Pilferage & Non-Delivery Clauses.

Claims to be payable in Bangkok. Insurance to be effect by Shipper.

Packing List in six-duplicate.

Certificate and List of Measurement &/or Weight in six-duplicate.

One copy of non-negotiable Bill of Lading

Two copies of non-negotiable Insurance Policy or Certificate.

Certificate of Origin in quadruplicate & all shipping documents to mention L/C No. 1254/67. Evidencing the shipment of 26 "SHIHLIN" Brand Transformers.

Details per Estimate No. ST67-049 dated August 10, 1989

Shipping Marks: Lopburi Co., Ltd.

Contract No. 63008-67-C-0150

CIF Bangkok from Taiwan to Bangkok partial shipments are not allowed. Transhipment is not allowed. Shipment(s) must be made not later than Nov. 10, 1981, and this Letter of Credit remains in force till Nov. 20, 1981.

We hereby agree with the drawers, endorsers, and bona fide holders of the drafts(s) drawn under and in compliance with terms of this Credit that such drafts shall be duly honoured upon due presentation and delivery of documents as specified, if drawn and negotiated within the valid date of this Credit.

The amount of any draft drawn under this Credit must be endorsed by the negotiating bank on the reverse hereof and the presentation of each draft, if negotiated, shall be a warranty by the negotiating bank that such endorsement has been made.

SPECIAL INSTRUCTIONS:

Negotiations under this Credit are restricted to Bank of Taiwan, Taipei, Taiwan.

Notes: Documents must conform strictly with the terms of the Credit. If you are unable to comply with its terms, please communicate with your customers promptly with a view to having the condition changed. NOTE CAREFULLY THE DESCRIPTION OF THE MERCHANDISE MENTIONED ABOVE AS IT MUST BE DESCRIBED IN EXACTLY THAT FORM IN YOUR DOCUMENTS.

Original and duplicate shipping documents to be sent by first airmail, remaining by subsequent airmail.

Yours faithfully,

for THAI DEVELOPMENT BANK, LTD.

信 用 狀（例六）

BANK OF AMERICA

IRREVOCABLE LETTER OF CREDIT

P. O. BOX 97, SINGAPORE SEPT. 3, 1989

Tang Eng Iron Works Co., Ltd.
65 Kuwan Chien Road
TAIPEI, TAIWAN

ALL DRAFTS & DOCUMENTS
DRAWN MUST BE MARKED:
DRAWN UNDER "B OF A"
CREDIT No. /4306

GENTLEMEN:

WE HEREBY ESTABLISH OUR IRREVOCABLE LETTER OF CREDIT IN YOUR FAVOR FOR ACCOUNT OF Hong Seh Co. Ltd. 177/179 Beach Road, Singapore UP TO THE AGGREGATE AMOUNT OF United States Dollars One Thousand Five Hundred Eighty Seven Only (US$1,587) AVAILABLE BY YOUR DRAFTS DRAWN AT

sight On accountee

AND ACCOMPANIED BY DOCUMENTS

SPECIFIED BELOW COVERING FULL (C.I.F.) INVOICE VALUE OF MERCHANDISE TO BE DESCRIBED IN INVOICE AS:

(400) Four Hundred Cases Bright Iron Wire Nails, countersunk, checkered head, diamond point in wooden cases of 20 kgs. net, 25 kgs. gross. Shipment from Taipei to Singapore latest Dec. 2, 1989; partial shipments are permitted. Transhipment is prohibited. Documents required marked below are to be forwarded to us by airmail in one cover.

Commercial Invoice in quadruplicate.

Certificate of origin.

Full set plus one non-negotiable copy of Clean On Board Ocean Bill Of Lading to order of shipper. Blank endorsed. Marked "Freight Prepaid" notify "Buyer" and ourselves.

Packing list in duplicate.

Marine and War Insurance Policy or Certificate in duplicate.

Blank endorsed for 110% of invoice value covering the Institute Cargo Clauses (F. P. A.)

The amount and date of each negotiation must be endorsed on the back hereof by the negotiating bank.

We hereby agree with bona fide holders that all drafts drawn under and in compliance with the terms of this credit shall meet with due honor upon presentation and delivery of documents as specified to accounts if drawn and presented for negotiation on or before December 2, 1980. Provisions applicable to this credit.

"This credit is subject to the uniform customs and practice for documentary credit(1962 revision)International Chamber of Commerce, brochure. No. 222."

Yours faithfully,
Bank of America
National Trust and Savings Association
By____________By____________

信用狀(例七)

THE BANK OF TOKYO, LTD.

KOWLOON OFFICE

657/661 Nathan Road

Mongkok, Kowloon

Via Airmail HONG KONG, 11th February, 1989

To: Taipei Trading Co., Ltd.
Taipei, Taiwan

All Drafts Drawn must indicate our credit number

Through: The Bank of Taiwan, Taipei, Taiwan

Gentlemen:

We open irrevocable documentary letter of credit No. KWI 0051 in favor of Yourselves

For account of HONG KONG TRADING CO., LTD., HONG KONG up to an aggregate amount of HK$62,589.25(Hongkong Dollars Sixty-Two Thousand Five Hundred and Eighty-Nine and cents Twenty-Five only)available by beneficiaries' draft or drafts at 90d/s sight on accountee for 100% invoice value when accompanied by the following documents:

signed commercial invoice in triplicate

Marine insurance policy or certificate in duplicate endorsed in blank, for 110% of the invoice value in currency of draft. Insurance must cover the institute war clauses the institute cargo clauses (All Risks) and the institute strikes, riots and civil commotions clauses.

Full set of Clean on Board Ocean Bills of Lading made out to the order of The Bank of Tokyo Ltd. Kowloon, marked "Freight Prepaid" and notify accountee" evidencing shipment from Taiwan to Hong Kong Covering.

6,452½ lbs. Worsted Hosiery Yarn, 2/20mm Top Dyed 100% Wool, full detailed per order #536. Colors:

OP-125 3,452½ lbs.

OP-127 3,000 lbs.

@HK$9.70 per lbs. C. I. F. Hongkong

Partial shipments are allowed. Transhipment is not allowed.

Shipment must be effected during February, 1981

Drafts must be negotiated not later than 10th March, 1981

Special instruction and other documents required if any:

All drafts drawn under this credit must be marked "Payable with interest at 8% per annum, from date of draft until approximate arrival of proceeds in Taiwan."

Unless otherwise instructed, the negotiating bank in reimbursement of their payment is authorized to draw sight draft on our office accompanied by their certificate certifying that all credit terms have been complied with and all documents have been airmailed to us in two consecutive lots.

We hereby engage with drawers, endorsers, and bona fide holders of all drafts drawn under and in compliance with the terms of this credit that such drafts will be duly honored.

Unless otherwise instructed this credit is subject to the uniform customs and practice for commercial documentary credits fixed by the international chamber of commerce (1962 Revision)

Negotiation under this credit is restricted to The Bank of Taiwan.
In reimbursement, we shall credit your account with our Hong Kong Office upon receipt of documents.
The Negotiating Bank must forward all documents to us by airmail.

Very truly yours,
For The Bank of Tokyo, Ltd.
Kowloon Office

有關信用狀的幾點說明

信用狀裏有幾個專用術語，必須略加解釋:

(1) 確認(Confirmed)　信用狀上如有了 "Confirmed" 一個字,

信用狀（例八）

Bangkok Bank Limited
TAIPEI BRANCH
(INCORPORATED IN THAILAND)
121, SUNG CHIANG RD, TAIPEI, TAIWAN
TEL: 5073275 TLX: 11289
SWIFT: BKKB TWTP

☐ This is the confirmation of our credit opened today by Telex/Swift

IRREVOCABLE DOCUMENTARY CREDIT

OUR REFERENCE NUMBER (PLEASE ALWAYS QUOTE)	DATE OF ISSUE
BBL-19761	January 15, 1989
DATE OF EXPIRY: March 10, 1989 IN THE COUNTRY OF THE BENEFICIARY	DOCUMENTS TO BE PRESENTED WITHIN DAYS AFTER THE DATE OF ISSUANCE OF THE TRANSPORT DOCUMENT(S) BUT WITHIN THE VALIDITY OF THE CREDIT

APPLICANT	BENEFICIARY
Taiwan Manufacturing Co. 456 Nanking East Road, Sec. 2 Taipei, Taiwan	Oriental Development Co. 3026 La Plata Avenue Hacienda Heights, New York
ADVISING BANK: Bangkok Bank New York	AMOUNT: US$1,870,000

PARTIAL SHIPMENTS	TRANSHIPMENT
Not allowed	Allowed

Shipment from : Keelung
for transpotation to : New York
latest shipment date : March 1, 1989

Credit available with
☒ by negotiation ☐ by acceptance
☐ by sight payment ☐ by deferred payment
against presentation of the documents detailed herein
☒ and of beneficiary's draft(s) at
drawn on

DOCUMENTS REQUIRED AT LEAST IN SET OF TWO IF NOT SPECIFIED

☒ Signed Commercial Invoice(s) in 2 copies indicating the Import Permit(s) number UEIP-4468

☒ Full set of CLEAN ON BOARD Ocean Bills of Lading and two non-negotiable copies made out to order of Bangkok Bank Ltd. marked: Freight ☒ prepaid ☐ to collect and notify applicant.

☒ Air Waybills consigned to Bangkok Bank Ltd. marked: Freight ☒ prepaid ☐ to collect and notify applicant.

☒ Insurance Policies or Certificates endorsed in blank for full invoice value plus 10 % with claims payable in Taiwan, covering ☒ Air Risks, ☐ Marine Risks, War Risks, as per Institute Cargo Clauses

☐ Packing Lists in 3 copies.

Special conditions:
☒ All banking charges outside Taiwan are for account of beneficiary.

We hereby issue this Documentary Credit in your favour. It is subject to the Uniform Customs and Practice for Documentary Credits currently in force and engages us in accordance with the terms thereof.

INSTRUCTIONS TO THE NEGOTIATING BANK:
The amount of each drawing must be endorsed on the reverse hereof. All documents are to be despatched to the issuing bank in two sets by consecutive registered airmails.
IN REIMBURSEMENT:

Yours faithfully,
For BANGKOK BANK LIMITED, TAIPEI BRANCH

Authorized Signature(s)
This Documentary Credit consists of 1 signed page(s).

Advising bank's notification

Place, date, name and signature of the advising bank

NO. IMP 19×88×6×1000

出口商就有了一種保障。因為銀行開發這一張以進口商為付款人的滙票 (Draft) 時，他們保證出口商，如果條件符合一定會收到貨款的。

(2) **不可撤銷** (Irrevocable) 這是說明銀行所開給出口商的信用狀，入口商或銀行都不能單方面將這個信用狀撤銷。出口商有了這種信用狀，便可高枕無憂了。

(3) **無追索權** (Without recourse) 這是對出口商進一步的保證。也就是說，出口商將滙票向銀行洽兌以後不管什麼事情發生，再也不能向出口商索回貨款了。假設上例臺灣的通用貿易公司 (General Trading Co.) 賣給西德的環球進口公司 (Universal Importers) 蘆筍六十噸，通用貿易公司憑信用狀滙票向臺灣銀行取得貨款美金三千八百元。臺灣銀行隨即將滙票及各項單據寄交西德聯行漢堡商業銀行收回墊款。可是在這期間，西德的環球進口公司倒閉了，貨價又大跌，漢堡商業銀行不但無法向環球進口公司要求付款贖單，就是拍賣和追保，也不能彌補所付還給臺灣銀行的款項。漢堡商業銀行只有自己承受損失，不能再向臺灣的出口商追索。所以信用狀上有了 Without recourse 二字，出口商便無後顧之憂。

(4) **發票全部貨款** (Full invoice value) 雖然進口商在信用狀上規定出口商可以開滙票收取（假定）五千美元的貨款。但這是指（假定）一萬碼棉布的貨價。倘使交運的實際數量比一萬碼少，他們所開出的滙票，當然不足五千美元。那麼，滙票的金額以什麼為根據呢？現在信用狀上指明了以發票的實際貨款作滙票的金額，這樣便明確多了。

(5) **清潔提貨單** (Clean bill of lading) 提貨單是由輪船公司簽發的。如果輪船公司在收到貨物時，發現包裝損壞，或木箱有翻釘的痕跡，或其他不妥的情形時，他們一定會在提貨單上加上這樣的"附註"。有這樣"附註"的提貨單就是"不清潔"的提貨單了。所以進口商在信用狀上一定會要求提貨單是“清潔”的，這自然是十分合理的。

進口商請求銀行開發信用狀時，銀行一定要向申請人收取一筆保證金（Margin）。因爲銀行是通知聯行替進口商先付款給出口商的，必須在貨物運到後，進口商才付款給銀行。爲了保證進口商一定可以贖單取貨，銀行照例要收一筆保證金作爲保障。至於保證金的多寡並無硬性規定。普通爲全部金額的百分之二十五。但主要的還是看貨物的性質以及銀行對進口商的信用而定。假定是化學原料或西藥，這是市價漲落較大的貨物，銀行所收的保證金比率一定高。對新創辦的公司或資金不大的公司，銀行也會收高些的保證金。所以預付保證金一項，在沒有開信用狀時，就必須和銀行商妥。

(6) 滙票(Draft or Bill of Exchange) 按照規定的手續，出口商在憑信用狀及各項單據，向所在地的銀行洽取貨款，應附一張滙票。這張滙票連同“跟單”（即各項單據之術語）一併交給銀行，由銀行寄交進口商所在地的聯行轉給進口商。進口商即照滙票上所列的金額付給聯行。如果是即期的（Sight Draft），那麼，應立即見票照付。如果是遠期的（三十天或六十天或九十天），可延遲付款。這就是所謂“押滙”。爲詳細說明起見，特舉例如下：

假定香港的一家貿易商向臺灣的義泰機器公司買了一批價機器，價值是五千美元。這家進口商請香港的恒生銀行開發信用狀，恒生銀行根據申請書，透過臺灣第一商業銀行，向義泰公司開出了信用狀。第一商業銀行接到恒生銀行的信用狀，立即轉給義泰公司。義泰公司將各項單據準備齊後，附上一張滙票（式樣如圖甲）：

Bill of Exchange

Draft No. K-18976

Exchange for US$5,000.00 — Taipei, Taiwan May 1, 1989

At sight of this FIRST of Exchange (Second of the same tenor and date being unpaid) Pay to the order of

FIRST COMMERCIAL BANK

The sum of US DOLLARS FIVE THOUSAND ONLY.

Value received

Drawn under Letter of Credit No. KM-75-02315 dated April 20, 1981

Issued by HENG SENG BANK, HONG KONG

To WATSON TRADING COMPANY, HONG KONG

YI TAI MACHINERY CO. LTD.

John Lee

Manager

69. 5. 3,000本(2×50)外出5 (2-1) (21×12公分)

(圖 甲)

這張滙票是第一商業銀行印好，交出口商義泰公司填寫的。（滙票號數由銀行填入）。滙票的背面（見圖乙）則由第一商業銀行的負責人簽字，也就是“背書”(Endorse)。有了第一商業銀行的背書，這張滙票就可以自由“讓購”(Negotiate)了。香港恒生銀行收到滙票及

Pay to the order of

ANY BANK, BANKER OR TRUST CO.

FIRST COMMERCIAL BANK

Alexander C. Cheng

Authorized signature

(圖 乙)

單據後卽通知進口商備款贖單。進口商付款後就可向輪船公司提貨，結束了這批交易。

上面圖甲裏有一句話 "this first of Exchange (Second of the same tenor and date being unpaid) 必須加以解釋。爲了防備滙票及跟單郵遞萬一遺失，銀行特通知出口商準備同樣的跟單兩份，滙票也分爲兩張，不在同一天寄出。圖甲滙票上證明這是第一張（第二張同案同日的另一張未付）。第二張則註明這是第二張（第一張同案同日的另一張未付）。倘使兩張都收到了，第二張就無效。萬一第一張遺失，可憑第二張付款，如此便十分妥當了。

再說，滙票上的"Issued by"是指開發信用狀的銀行。"TO"是指受益人。受益人可能就是開發滙票的銀行，也可能是另一家銀行或公司，這是在信用狀上所指定的。

習　題

(一) 試述信用狀的定義。

(二) 信用狀內中包含何種項目?

(三) 簡述申請進口信用狀的手續。

(四) 信用狀應該如何核對?

(五) 爲什麼信用狀裏規定要保險百分之一百一十 (110%)?

(六) 試述押滙及跟單的意義。

第六章　商用電報

在商業上，除了絕大部份利用書信接洽和進行交易外，有時爲了爭取時間，也須利用電報。尤其是國際貿易，使用電報的時候更多。現在將有關國際電報的各項，分列如下：

第一節　國際電報的種類

（甲）書信電報（Letter Telegram）簡寫爲，“LT”

書信電報適用於無特別重要時間性的明語電報。照尋常電報價目減半收費，最少二十二字起算。並須在收報人姓名住址前加註“LT”標識，也作一字計算。投遞辦法，是按照發報國家地區，分別於每日上午八時起及下午二時起開始投送。

按照我國電信局的規定，有些地區是不適用書信電報的。例如香港和澳門，就不能收發書信電報。其他尙有法國、比利時、象牙海岸、瑞典等地。

爲了節省電報費，一般商號，多用書信電報。但如果字數很少，那麼可能拍發尋常電報，還便宜些。例如拍往日本的書信電報，起碼收費爲新臺幣128.70元，如果這封電報只有十個字，那麼，按尋常電報每字11.70元計算，僅需117元，反而比較合算。

（乙）尋常電報(Ordinary Telegram)

電文可用明語或密語， 或兩種混合書寫。 報費最少以七個字起算。

（丙）加急電報(Urgent Telegram)

加急電報照尋常電報價目加倍收費，以七個字起算。享有優先傳遞及投送的權利。但須在收報人姓名住址前加註一個“Urgent”標識，並做一字計算。

第二節　計算字數方法

（甲）按照英文標準字典所載，單字凡不超過十五個字母的，作一個字計算。每超過十五個字母的，加計一字。例如:
Incomprehensible 共十六個字母，應作兩字計費。

代表貨幣的符號，在電報上通常都把它們拼出來，因爲符號也算一字，並無分別。所以 Pound 可代替 £，Dollar 可代替 $。其他單獨的標點符號，例如（，），（·），（：），（；），（？）等都作一字計算，所以在電文中應該省免。

（乙）密語、商務標識、數目字、連續的標點符號、縮寫字等，都以每五個字碼作一字計費。字母、數目字及符號混合組成的字，計費方法，按國際電信局電報規則的規定辦理。

（丙）明語字組的不規則湊合，在尋常電內應按每五個字母作一字計費。在書信電內需按原有的意義，按字收費。

下列各字，都被認爲一個字:

15×35　2×4×5　50-60　45%　4PM　11AM

REURS　RETEL　RELET

下列各字，却要按兩個字計費：193+15　105.95　38.83%

第三節　書寫電報方法

（甲）國際電報可用明語，或密語，或兩者混合書寫。明語限用中文（四碼）英文、法文、及羅馬字拼音日文。

（乙）書信電報及新聞電報，不得含有暗語或密語。

（丙）電報最好用打字，每字之間應留有明晰間隔。如用手寫，必須用大寫方體字母，以免發生錯誤。

（丁）收報人及發報人姓名住址，必須詳細寫明。發報人姓名住址，如果不要拍發，需寫在電文欄紅線以下。舉例如下：

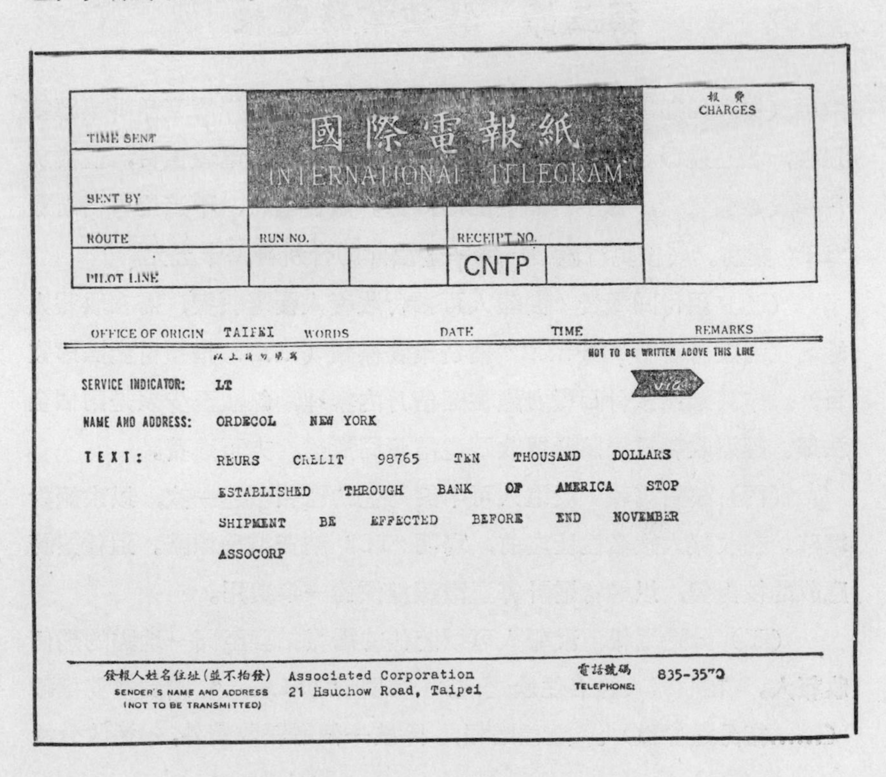

國際電報紙
INTERNATIONAL TELEGRAM

TIME SENT
SENT BY
ROUTE　RUN NO.　RECEIPT NO.
PILOT LINE　CNTP
報費 CHARGES

OFFICE OF ORIGIN　TAIPEI　WORDS　DATE　TIME　REMARKS

以上請勿填寫
NOT TO BE WRITTEN ABOVE THIS LINE

SERVICE INDICATOR:　LT　　VIA

NAME AND ADDRESS:　ORDECOL　NEW YORK

TEXT:　REURS　CREDIT　98765　TEN　THOUSAND　DOLLARS
ESTABLISHED　THROUGH　BANK　OF　AMERICA　STOP
SHIPMENT　BE　EFFECTED　BEFORE　END　NOVEMBER
ASSOCORP

發報人姓名住址（並不拍發）
SENDER'S NAME AND ADDRESS (NOT TO BE TRANSMITTED)
Associated Corporation
21 Hsuchow Road, Taipei
電話號碼 TELEPHONE:　835-3570

第四節 如何交發電報

（甲）國際電報可交國際電信局營業處，以及當地各電信局營業處拍發。

（乙）除下列情形外，任何報費，都要在發報時付現金：

(1) 記帳發電，免付現金的。

(2) 以足額的預付回報費，憑單抵付的。

(3) 經事先洽妥，由收報人或第三者付費的。

第五節 特種業務電報

（甲）船舶電報　船舶在海洋中航行，經由海岸電臺呼叫或應接以後，相互通報。報費按拍發至海岸電臺所在地的電報價目，並附加海岸及船舶報費，每字一個金法郎計算。這種電報，不適用書信電報"LT"業務。（按現行滙率，每一金法郎約合新臺幣十五元。）

（乙）預付回報費　發報人可預付收報人覆電報費，需在收報人姓名住址之前書明"RP......"納費業務標識（.........指預付金法郎數目）。申請退還預付回報費應在四個月內辦理，餘額至少須是兩個金法郎。經過收報電局將憑單收回並電覆同意後，方可退費。

（丙）校對電報　發電人可申請將他的電報覆述一次，以求絕對準確。在收報人姓名住址之前，寫明"TC"納費業務標識。這種業務應收的校對費，以尋常電計算這電報總價的一半費用。

（丁）分送電報　發報人可以將他的電報拍發至同一地點的幾個收報人。在收報人姓名住址之前，書明"TM......"納費業務標識（......指分送份數）。這種電報，除按一個電報收費外，並收分抄費，每份在五十個字以內的，收一個金法郎，每超過五十字的，加收五十生丁(即半個金法郎)。

（戊）新聞電報　新聞電報的電文，必須是新聞的報導，用來刊登新聞報紙，定期刊物，或無線電臺及電視廣播。最少十四個字起算，可享受特別減價優待。但發報人需具有新聞記者身份，經電報局認可後方可拍發。

（己）美國國際航海衞星通訊電報　美國國際航海衞星電報（限尋常電），除按現行中美電報價目計收外，每字另加航海衞星電路費每字新臺幣十五元三角。

第六節　如何簡省電文

電報收費，旣然是以字計算，那麼，自然以字愈少愈好。凡是能夠簡化的應加簡化，能省略的就應省略。例如With reference to your telegram 五個字，用 RETEL 來代替，只須一個字。Letter of credit 三個字，用 L/C 就是一個字。書信電報最好不要超過廿二字的起碼收費規定。因此電文不能像書信一樣，講求文法或多用客套，應以簡單明瞭爲主。在信函中常用的 Please 及 Thank you，均應省略。下面便是幾個例子：

(1) Your immediate reply is requested......*Reply immediately*

(2) We await your instructions......*Instruct immediately*

(3) Please give us a full account......*Inform fully*

(4) We cannot comply to your request ......*Regret*

(5) Your cable (letter) of 25th......*Yours 25th*

(6) Please refer to our letter of 23rd......*Ours 23rd*

(7) We will write you......*Write*

(8) We are airmailing you the details......*Airmailing details*

(9) Owing to the fact that......*Owing*

(10) The quality of goods is not satisfactory......*Quality unsatisfactory*

(11) We have not received your letter of 24th......*Yours 24th unreceived*

(12) We regret that they refuse to effect payment......*Regret their nonpayment*

(13) We appreciate your kindness......*Appreciate*

(14) With reference to your letter of 21st......*Relet 21st*

(15) With reference to your telegram of 17th......*Retel 17th*

(16) We apologize the delay of our shipment ...... *Apologize delayed shipment*

(17) We shall open Letter of Credit immediately......*Open L/C immediately*

(18) Please open Letter of Credit without delay......*Expedite L/C*

(19) We regret we do not have the goods at present ......*Goods now unavailable*

(20) Please settle the payment without delay......*Pay soonest*

另一個省字的辦法，是利用"電報掛號"來替代收報人及發報人的姓名地址。照國際電報局的規定，"電報掛號"以十個字母爲限，由申請人自定，並無限制，但不能和他家重複。例如臺北市南京東路二段五十四號三樓統一棉花公司，只消用"UNICOTCORP"一個字就可代表了，豈不是省去了不少的電報費。凡在電報局登記付費以後的"掛號"，電報局於收到電報後可以立卽按址迅速送到。

還有一種簡化電文的方法，是使用密碼字在電文裏面。密碼分爲公共密碼和私人密碼兩種。經大家公認的公共密碼書，計有阿克米商品及用語密碼(ACME Commodities and Phrases Code)；本特萊第二組合密碼 BENTLEY II (Bentley's Second Phrase Code)；愛皮西環球商業電碼 ABC (ABC Universal Commercial Telegraph Code)；彼得遜國際密碼 PETERSON (Peterson International Code) 等。其中以 ACME 爲一般商業所採用，銀行界則普遍採用 Peterson International Code。此外尙有在國際貿易上應付買賣雙方特殊關係、環境、經營範圍而私下所編的的私人密碼。這種密碼自然比公共密碼來得更爲合適，更能保守秘密。

使用密碼雖可省錢，但也有缺點。因爲萬一密碼在收到時，其中有幾個字母的錯誤，便會使整個電報的意義不易明瞭。

第七節 國際電報費率

我國國際電信局訂有國際電信價目表，目前由臺灣可以直接和間接通報的地區，共達二百一十八處。下面是幾個主要地區的收費表：

目的地	尋常電報	書信電報	
	每字（新臺幣）	起碼廿二字(新臺幣)	每加一字（新臺幣）
加拿大	24.80	273.00	12.40
德國	37.20	409.50	18.60
香港	6.60	—	—
意大利	37.20	409.50	18.60
日本	14.90	164.00	7.45
韓國	14.90	164.00	7.45
澳門	6.60	—	—
菲列賓	14.90	—	—
泰國	24.80	273.00	12.40
英國	37.20	409.50	18.60
美國	24.80	273.00	12.40
星加坡	24.80	273.00	12.40

下面是明碼和密碼的電報舉例，並計算應付的電報費：

(1) 明碼書信電報

LT UNIMPORINC HAMBURG WESTGERMANY
REYUR INQUIRY 3RD OFFER FIRM VALIDITY TWO WEEKS 20 METRIC TONS EACH TIP CENTER END CUTS US CENTS 75.00 71.00 44.00 PERKG CANDF HAMBURG SEPTEMBER SHIPMENT GENETRACO

這個電報連同"LT"共計三十一字，照上表拍至德國的書信電報，最初廿二字是新臺幣四百〇九元五角，每加一字，收費十八元六角，因此應付的電報費是新臺幣五百七十六元九角。

(2) 明碼尋常電報

TAKSANTRAD HONG KONG
RELET 17TH OFFER 10,000 METRIC TONS SWC SUGAR US DOLLARS 105 PER TON FAS KAOHSIUNG DELIVERY FOUR WEEKS AFTER RECEIPT LC

TAISUGCORP

這封電報共二十四字。因爲國際電信局規定香港不適用書信電報，所以只能按尋常電報拍發，應付費每字六元六角，共計新臺幣一百五十八元四角。尋常電前面不需要加 ORDINARY 字樣。

(3) 加急明碼電報

URGENT HONKONCONT MACAU
RE ENQUIRY JULY 30TH QUOTE ONE MILLION SQUARE FEET 3PLY LUAN PLYWOOD US DOLLARS 29.50 PER 1000 SQUARE FEET FOB KAOHSIUNG OCTOBER SHIP-

MENT PLEASE CABLE REPLY IF ACCEPTABLE

UNIENTERCO

這份電報，連同"URGENT"共三十二字，每字新臺幣十三元二角，應付報費新臺幣四百二十二元四角。

(4) 尋常密碼電報

TEXVOTCORP DALLAS TEXAS USA
IGTYM KEELUNG KOEOD UWAMY 1/16" OAMOX
NYSMO MIHOM　FORTEXTILE

按照阿克美密碼 (ACME Commodites and Phrases Code)譯文如下:

IGTYM	Please make best possible firm offer CIF
KEELUNG	
KOEOD	1000 bales
UWAMY	Middling Uplands
1/16"	
OAMOX	Shipment in three months
NYSMO	First shipment during October
MIHOM	Reply immediately
FORTEXTILE	

以上共計十三字，按表每字拍至美國收費新臺幣二十四元八角，共計新臺幣三百二十二元四角。

(5) 加急密碼電報

URGENT GENERIMPOR NEWYORK
FOXZU IHKAY NEWYORK YESEA POUVJ AWCLO 58

OCHFU MIGTS TAIPINCORP

譯文如下：

FOXZU	Referring to your letter of 30th instant
IHKAY	We offer firm CIF subject to reply received here by the 4th
New York	
YESEA	Standard sliced pineapple No. 1 special 4 dozens to case
POUVJ	US$4.80
AWCLO	(Per case) of......tins
58	
OCHFU	Shipment per steamer during November
MIGTS	Prompt reply required
TAIPINCORP	

這封電報連"URGENT"也是十三個字，照表應收每字新臺幣四十九元六角，共計新臺幣六百四十四元八角。

第八節　電報詢價和答覆實例

下面是明語電報詢價及答覆的例子：

(1) 西德來電（詢蘆筍價）

July 8, 1989

LT

GENTRACO TAIPEI TAIWAN

RE CANNED ASPARAGUS PLEASE CABLE QUOTE FOB

KEELUNG CANDF HAMBURG 20 M/TS EACH TIPCUTS
CENTERCUTS ENDCUTS EARLIEST SHIPMENT

UNIMPINC

(2) 臺灣覆電（報蘆筍價）

July 11, 1989

LT

UNIMPINC HAMBURG WESTGERMANY

REURTEL 8TH OFFER 20 M/TS EACH TIPSCUTS CENTERCUTS ENDCUTS USCENTS 69.00 66.00 39.00 PERKG FOB KEELUNG NET OR 75.00 71.00 44.00 PERKG CANDF HAMBURG NET SHIPMENT ONE MONTH AFTER RECEIPT L/C

CENTRACO

(3) 香港來電（詢特級白砂糖價）

July 17, 1989

TAISUGCORP TAIPEI TAIWAN

INTERESTED 10000 M/TS SWC SUGAR FOR INDONESIA PLEASE QUOTE FAS KAOHSIUNG OR CANDF JAKARTA NET PROMPT SHIPMENT AWAITING IMMEDIATE REPLY

TAXATRACO

(4) 臺灣覆電（報特級白砂糖價）

July 8, 1989

TAXATRACO HONGKONG

YC 15TH OFFER FIRM VALID UNTIL 31ST 10000M/T SWC SUGAR USDOLLARS 105.00 PER M/T FAS KAOHSIUNG NET STOP AWAITING EARLIEST REPLY

TAISUGCORP

(5) 西德來電（詢混合紡織品價）

July 15, 1989

LT

UNITEXCO TAIPEI TAIWAN

PLEASE OFFER ROCKBOTTOM PRICES C&FC2 AMSTERDAM 50000 YDS EACH BLENDED FABRICS 65% POLYESTER 35% COTTON 88×75 130×75 110×70 ALL 45/45 AND 57" PROMPT DELIVERY

HERSINCO

(6) 臺灣覆電（報混合紡織品價）

July 18, 1989

LT

HERSINCO HAMBURG WESTGERMANY

REURTEL 15TH QUOTE BLENDED FABRICS 65% POLYESTER 35% COTTON 50000 YDS EACH 88×75 110×75 ALL 45/45 57" USCENTS 35.60 41.70 36.00 PERYARD CIF NET HAMBURG STOP AWAITING CABLE CONFIRMATION

UNITEXCO

(7) 香港來電（詢柳安三夾板價）

July 11, 1989

LT

UNIENTCO TAIPEI TAIWAN

PLEASE OFFER BEST PRICE ONE MILLION SQUARE FEET 1/8″×3′×6′ 3-PLY LAUAN PLYWOOD TYPE III FAS KAOHSIUNG EARLIEST SHIPMENT

HONKOWCONT

(8) 臺灣覆電（報柳安三夾板價）

July 14, 1989

HONKOWCONT HONGKONG

YC11 OFFER ONE MILLION SQUARE FEET 1/8″×3′×6′ 3PLY LAUAN PLYWOOD TYPE III USDOLRS 29. 50 PER 1000 SQUAREFEET FOB KAOHSIUNG NET OCTOBER SHIPMENT PLEASE CABLE CONFIRM IF ACCETABLE

UNIENTCO

(9) 英國來電（催答十日函中所詢事項）

February 12, 1989

LT

TEHSINGMIL TAIPEI TAIWAN

FURTHER OUR LETTER IOTH PLEASE TRY UTMOST SUBMITTING PRICES BOTH FOB CIF EARLIEST POSSI-BLE

ANGAFTRACO

(10)臺灣覆電（報毛線衫價）

February 18, 1989

LT

ANGAFTRACO LONDON ENGLAND

REURLET 10TH OFFER FIRM UNTIL 28TH ACRYLIC SWEATERS LARGE MEDIUM SMALL USDOLRS 12. 50 10. 50 9. 00 PER DOZEN FOBC2 KEELUNG 14. 00 12. 50 11. 50 CIFC2 DURBAN PLEASE CABLE CONFIRM

TEHSINGMIL

(11)臺灣去電（詢小麥價）

February 12, 1989

LT

AUSEXPTYCO MELBOURNE AUSTRALIA

PLEASE QUOTE ROCKBOTTOM CIF KEELUNG 20000 MTS NO3 1980 CROP WHEAT PROMPT SHIPMENT AWAITING IMMDEIATE REPLY

SUNMOTRACO

(12)澳洲覆電（報小麥價）

February 19, 1989

LT

SUNMOTRACO TAIPEI TAIWAN

YC12 OFFER FIRM UNTIL 28TH 20000 MTS WHEAT NO3 1980 CROP AUSDOLRS 252 PERTON CIF KEELUNG

NET PLEASE CABLE CONFIRM IF ACCEPTABLE

AUSEXPTYCO

第九節 國際電報交換(TELEX)

(一) 國際電報交換是什麼?

許多大公司商號或銀行,爲了避免拍發電報,要到國際電信局的麻煩,並盼立即得到對方的回音,常在自己的辦公室裏裝置一種叫做 Teleprinter 或 Teletype 的機器,向電信局掛號之後,就可以和對方彼此交換電信。這種電信,稱爲 TELEX 或 TLX,是 Teletypewriter Exchange 的簡稱。每一個參加的行號,都有一個編號,就是一般人所稱的 Telex Number。

國際電報交換是利用一個電報交換機 (TELEX MACHINE),將電報的文字變爲信號 (SIGNAL),使透過各種複雜的通信系統,送到國外裝有同樣設備的被呼用戶,以提供一項最迅速的通訊服務。

電報交換機是用來將電報的文字,傳遞到國外你所需要通信的用戶。當你在自己的電報交換機上打出文字時,在國外你所通信的用戶的電報交換機上,同時也自動打出同樣的文字。

國際電報交換業務和國際電話業務性質,大致相同。只是使用的機器,是以電報交換機代替電話機,並以電報文字代替電話聲音。

(二) 國際電報交換的優點

這種新發明的國際通信方法,對於商業上有很大的幫助。因爲它有下面的好處:

(1) 憑電傳打字機收發電報,自己有可靠的紀錄,作必要時的查考。

(2) 可以立卽得到對方的答覆。

(3) 收報人不在，自動打字機照常可以收電，它會將電文記錄在紙上。對於時差較大的外國通信，更爲便利。

(4) 可以送出簡單的統計表。

(5) 直接在辦公室用電報交談，和用電話通話相同，十分方便。

(6) 無須將電報送往電信局拍發。

(7) 電信局也免除派人送報的麻煩。

(8) 按通信的時間計費（至少一分鐘或三分鐘），比較普通電報按字計費，低廉很多。

需要利用TELEX的用戶，要事先向國際電信局申請裝設專線，和五個單位啓閉式電傳打字機全套，卽由國際電信局列入電報交換用戶，並編列號碼。（像裝電話後電信局給用戶一個號碼一樣）以後就可直接或經掛號手續後和國外 TELEX 用戶直接通報。反過來說，國外用戶也可以直接或經過掛號後，和臺北的用戶直接通報。

另一好處，是用戶裝設了專線和電傳自動打字機，除了可以掛接 TELEX 外，還可以拍發對方未裝 TELEX 機的一般性的國際電報。過去用戶要發國際電報時，必須親自或派人將電報送到國際電信局，現在只要利用機器，先叫通國際電信局的專線臺，將電報傳過去。這種業務，叫做專線業務 Private Tieline Service 簡稱 PTL。同時國外來報，也可以利用此種專線將來報直接傳給用戶，不再用人力投送，可稱便捷之至。

（三）電報交換呼叫接線的分類

國際電報交換呼叫的接線種類，要看各被呼用戶所在國（或地區）的交換設備情形而定。大致可分爲三類:

(1) 人工交換接線(MANUAL OPERATION)呼叫用戶先將被呼用戶的電報交換呼叫號碼及所在地名，向當地電信機關電報交換席值機員掛號後， 由值機員請被呼用戶之交換席值機員， 將被呼用戶接出，然後互相通報。

(2) 半自動交換接線(SEMI-AUTOMATIC CONNECTION) 呼叫用戶先將被呼用戶的電報交換呼叫號碼及所在地名，向當地電信機關交換席值機員掛號後，由值機員直接將被呼用戶接出，然後互相通報。

(3) 全自動交換接線(FULLY AUTOMATIC CONNECTION)呼叫用戶可直接選接國外被呼用戶的電報交換呼叫號碼後，直接將被呼用戶接出通報。

(四) 國際電報交換的設備及使用說明

TELEX 用戶裝用的機件是一整套的收發兩用電動錄報機。 它包括：

(1) 頁式電傳打字機一架。

(2) 紙條鑿孔機一架。

(3) 紙條發送器一個。

(4) 線路控制器一組。

(5) 操作用樞紐數個。

(見附圖及使用法說明)

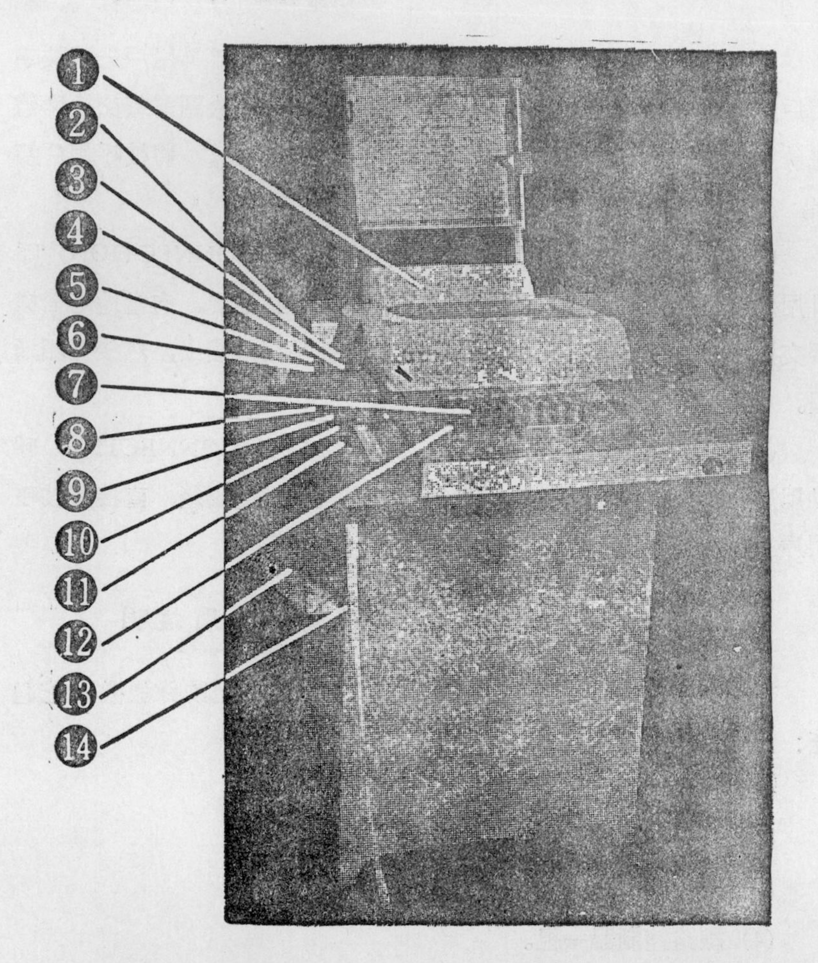

1. 捲筒紙(PAPER ROLL)

   供來去電報自錄報底之用。

2. 油紙條(TAPE ROLL)。

   用戶可將去電先拍鑿於此紙條上，以便與國外用戶接通電路後，將此紙條透過五單位電碼發報機而將去電傳送國外用戶。

3. 鍵送電鈕(TAPE OFF PUSHBUTTON)

   將此鈕按下，不論電路接通與否，用戶若於鍵盤上按動電鍵時，均僅有紙頁自錄抄件，而無紙條產生。

4. 紙條鑿孔電鈕(TAPE ON PUSHBUTTON)

按下線路控制開關（圖12）於內線位置，再按下此鈕，即可拍鑿去報紙條。同時，電傳打字機上亦有字跡打出，但無訊號發出，此乃僅供用戶自己拍鑿紙條之用。但須注意：倘線路控制開關放置於線路(LINE)位置，且於線路接通後（即被呼用戶之自動呼叫號碼及回應電碼出現於用戶之電報交換機上後）用戶若按下此鈕，再按動鍵盤上任一電鍵，此時即可將該電鍵之訊號直接發送被呼用戶，紙條亦同時被拍鑿爲五單位電碼之各樣排列不同之小孔，用戶可將此紙條暫時保留，以備被呼用戶要求重發時，即可利用此紙條由紙條發報器(T.D.)重發一次，以省人力。

5. 紙條釋放鈕(TAPE RELEASE PUSHBUTTON)

當紙條穿過鑿孔機而遇困難時，即可按下此鈕，將紙條自由調整（前進或後退）使之靈活運用。

6. 倒退電鈕(TAPE BACK SPACE PUSHBUTTON)

此鈕每按動一次，即可將已鑿孔之紙條退回一字，作爲校正錯誤之用。

7. 詢問鍵(WHO ARE YOU)

此鍵位於"D"字之上行，當需用時，必須先按數目鍵（FIGS KEY），再按D字鍵，藉以索取被呼用戶之自動呼叫號碼及回應電碼（AUTOMATIC CALL NUMBER AND ANSWER-BACK CODE）故自動呼叫號碼及回應電碼電鈕(HERE IS)乃專爲回答此鍵之訊問而設。

8. 鑿孔機（PERFORATOR）：用以拍鑿去報紙條之用，如將紙條鑿孔電鈕④按下，於鍵盤上按動電鍵時，此機即開始將去報紙條鑿出五單位電碼同樣字母之各種不同位置之小孔，用以透過紙條發送器後，發送去報之用。

9. 紙條出口(TAPE OUT)

10. 紙條發報器(TAPE TRANSMITTER)

凡已鑿孔之紙條，透過此五單位電碼發報器時，因紙條上所鑿之五單位電碼小孔排列位置不同，而產生各種不同之五單位電碼訊號，發送給收報用戶。

11. 紙條發報開關：當電路接通後，將已拍鑿之紙條，放壓於紙條發報器上，將此開關推向發報開始 (START) 位置時，紙條便可自動前進，電報亦被發出。將此開關拉回發報停止 (STOP) 位置時，紙條即停止前進，電報亦停止發送。若將此開關再向後拉至釋放 (FREE) 位置時，紙條便可前後任意伸縮，隨意安排。但應注意：若線路接通後，如不利用紙條發報，僅於鍵盤上按動電鍵時，此開關不論放置任何位置，訊號均可直接發送至被呼用戶，故應特別小心。

12. 數目字鍵(FIGS)：電文內，任何數目字或各種標點符號，如需拍發，即應先按此鍵，再按所需之數目字或標點符號鍵。
13. 盛紙屑箱(WAD BOX)：去報紙條，經鑿孔後，所有脫落紙屑，均自然掉落此箱內，以保機內及工作環境之清潔美觀。
14. 已鑿孔電報紙條(PERFORATED TAPE)：
去報已被鑿孔爲五單位電碼之紙條。

因爲 TELEX 是兩用戶之間的直接通信，可以不必拘泥於電報的格式，也不需要報類和其他標識，雙方可以隨意自訂格式，以能達到相互通信的目的就可以了。下面便是一個例子：

TP500 TRADECO CLG 54321 TRADECO GENEVE 12 JUNE
MAY WE CALL YOUR ATTENTION TO OUR ACCOUNT OF LAST MONTH WHICH HAS A PAID BALANCE OF DOLLARS 12345. 10. IF YOU DID NOT RECEIVE AN ITEMIZED BILL OR IF THERE ARE ANY ITEMS YOU QUESTION PLEASE LET US KNOW AND YOUR REPLY WILL RECEIVE OUR PROMPT ATTENTION. MAY WE HEAR FROM YOU?
COL: 12345. 10 END OF CALL

TELEX 的確在商業上有很大的便利。唯一的缺點是因爲計費按時間而不按字數算，因此起稿人容易產生“無所謂”的心理，不再推敲電文。往往本來可以用簡潔的通信，變成冗長的文字。

（五）電報交換 *TELEX* 的費率

現在和臺北可以互通電報交換 TELEX 的，共有一百八十二個地區，一部份已開放全自動撥號，計有阿根廷、澳大利亞、奧地利、比利時、巴西、加拿大、丹麥、芬蘭、法國、德國、夏威夷、香港、意大利、日本、科威特、荷蘭、挪威、菲列賓、波多利柯、星加坡、西班牙、瑞典、瑞士、英國、美國、委內瑞拉等地。全自動撥號，通報費按一分鐘起算。半自動接線，通報費按三分鐘起算，有澳門、阿拉斯加、冰島、肯亞、洪都拉斯、海地、加彭、達荷美、中非共和國、查德、緬甸、波里維亞、尼泊爾、墨西哥、烏拉圭等地。下面是幾個主要地區的收費表（另加稅一成）：

發往地點	基本計費分鐘	每一分鐘通報費（新臺幣）	超過三分鐘每分鐘加收（新臺幣）
法國	1	152.00	—
德國	1	152.00	—
香港	1	101.00	—
意大利	1	203.00	—
日本	1	152.00	—
加拿大	1	152.00	—
瑞士	1	152.00	—
菲列賓	1	152.00	—
波多利柯	1	203.00	—
英國	1	152.00	—
美國	1	152.00	—
澳門	1	101.00	—
阿拉斯加	3	531.00	161.00
冰島	1	152.00	—

習題

(一) 試述國際電報的種類。

(二) 試述電報計算字數方法，並舉例說明之。

(三) 何謂特種業務電報?

(四) 電文應如何節省? 試舉例說明之。

(五) 試擬書信電報一通，向西德詢問原棉價格。

(六) 何謂國際電報交換(TELEX)?

(七) 試述國際電報交換(TELEX)的優點。

(八) 試述國際電報交換的收費方法。

第七章　商業合同

除了商業書信和商業電報而外，商業合同也是商用英文的一個重要部份。

商業合同的種類很多，本書所舉的實例，計包括下列各種：

(1) 買賣合同。

(2) 聘雇合同。

(3) 代理合同。

(4) 加工合同。

(5) 建築合同。

(6) 租屋合同。

(7) 抵押合同。

因爲合同是法律的根據，因此它們有一定的格式，文字方面，尤其涉及金錢部份，更不可有絲毫含糊。

倘使合同用兩種文字（例如中文和英文），那麼，必須說明在萬一發生疑義時，應以何種文字的解釋爲標準。

爲確保合同的法律效力，最好的辦法是辦理法院公證手續，或由律師作證。

第一節 買賣合同

（購買五百噸美棉）

AGREEMENT OF PURCHASE

This Agreement made this first day of November, 1989 between The International Transportation Trading Co., Ltd. of 20, Lane 18, Chienkuo North Road, Taipei, party of the first part, and The Ta Tung Textile Corp. of 1928 Po-ai Road, 2nd fl., Taipei, party of the second part.

Witnesseth:

That is, said party of the first part in consideration of the agreement of the said party of the second part, hereinafter contained, contracts and agrees to and with the said party of the second part, that he will deliver, in good and marketable condition, at the City of Taipei, during the month of April of next year, five hundred tons of American Cotton, in the following lots, and on the following specified terms, viz: —one hundred tons by the seventh of April, one hundred tons additional by the fourteenth of the month, one hundred tons more by the twenty-first, and the entire five hundred tons to be all delivered by the thirtieth of April.

And the said party of the second part, in consideration of the prompt fulfillment of this contract by the said party of the first part, contracts and agrees with the said party of the first part to pay for said cotton twenty thousand New Taiwan Dollars per ton for each ton as soon as delivered.

In case of failure of agreement by either of the parties hereto, it is hereby stipulated and agreed that the party so failing shall pay to the other four thousand New Taiwan Dollars per ton as fixed and settled damage.

In witness whereof we have hereunto set our hands the day and year first above written.

witness: The International Transportation Co., Ltd.

Beverly Lu Caroline Lee

(Manager)

Witness: Ta Tung Textile Corp.

Margaret Hsieh Vivian Huang

(Manager)

【譯　文】

購買合同

本合約於一九八九年十一月一日由臺北建國北路十八巷廿號國際運輸貿易公司（甲方）及臺北博愛路一九二八號二樓大同紡織公司（乙方）簽訂。

根據甲乙雙方所訂的合約，甲方應在明年四月交付上好可以出售的美國棉花五百噸，條件如下：四月七日交一百噸，四月十四日再交一百噸，四月廿一日又交一百噸，四月卅日將全數五百噸交足。

乙方爲履行合約，同意每噸付款新臺幣二萬元，於交貨時隨即付淸。

任何一方倘不遵守本合約，應賠償對方損失每噸新臺幣四千元。

爲昭信守起見，特於上開日期簽約爲憑。

證人：國際運輸公司

呂玲珍：李慧蘭

(經理)

證人：

謝學和　大同紡織公司

黃翠玉

(經理)

第二節 聘雇合同

(聘用工程師)

AGREEMENT OF EMPLOYMENT

.his Agreement made and entered on this first day of February 1989, between David Simon (hereinafter called the first party) and David Lee (hereinafter called the second party).

Whereas the first party is willing to employ the second party and the second party agrees to do as the first party's engineer in Los Angeles. It is hereby mutually agreed as follows, that is to say:

1) The employment is preliminarily for a period of three years from March 1, 1989 to February 28, 1992, exactly functioning from the date the second party arrives and he will return to his country when he has completed his 3 years service.

2) The second party's monthly salary is US$1,500.00 which shall be paid monthly in Los Angeles not to be delayed for any reason. The second party's housing shall be provided by the first party during the service period.

3) The second party shall be provided by the first Party with the air ticket from Taiwan to Los Angeles and from Los Angeles to Taiwan upon terminating the service.

4) The second party shall obey the law of the United States and do his responsibility as a citizen when he is there.

5) This agreement is made both in English literature and Chinese literature in triplicate, each of which, both English copy and Chinese

copy, shall be kept by either party and the Office of Public Notary, Taipei District Court.

The First Party:

David Simon

The Second Party:

David Lee

【譯 文】

聘雇合同

本合約於一九八九年二月一日由大衛西蒙（以下簡稱甲方）及李大維（以下簡稱乙方）簽訂。

甲方顧聘乙方，乙方也同意應聘爲甲方在洛杉磯的工程師。雙方協議條件如下：

(1) 聘約初訂爲期三年，自一九八九年三月一日至一九九二年二月廿八日，自乙方抵達日正式開始生效。乙方在完成三年工作後，應即返回本國。
(2) 乙方月薪爲美金一千五百元，由甲方在洛杉磯支付，不得以任何理由拖延。在服務期內，甲方供給乙方的住所。
(3) 乙方自臺灣至洛杉磯，以及在服務期滿，自洛杉磯回臺灣的飛機票，由甲方供給。
(4) 乙方在美國時應遵守美國法律，盡做公民的義務。
(5) 本合約中英文本各一式三份。雙方及臺北地方法院公證處各執中英文本一份。

甲方

大衛西蒙

乙方

李大維

第三節　代理合同

（代理經銷臺灣冷霜）

COMMERCIAL AGENCY AGREEMENT

Sincerely made this Agreement on the first day of April in the year of our Lord, one thousand nine hundred and eighty nine between Mr. Wu Tong-san, representative of Messrs. Tong Chi Industrial Corporation, No. 432, Kin Lo Road, Tainan City, Taiwan Province, Republic of China (Hereinafter called the first party) and Mr. C. Y. Cheng, representative of Messrs. Hwa Tai Trading Company, Ltd., No. 3425 A, Sanila Street, Manila, Philippines (Hereinafter called the second party).

Whereas the first party is willing to appoint the second party and the second party agrees to do as the first party's sole agent in the territory of Philippines for selling VIP Cold Cream one of the first party's products. It is hereby mutually agreed as follows, that is to say:

1. The first party appoints the second party as the first party's sole agent in the territory of the Philippines for selling VIP Cold Cream.
2. The first party shall supply the second party with the paste of VIP Cold Cream; the second party shall have it wrapped and disposed of with the trade mark and labels copied after the original ones.
3. The second party shall sell monthly not less than 5,000 dozens.
4. This agreement shall be cancelled immediately at any time in case the second party fails to sell up to the agreed quantity for six months.
5. During the term of this agreement, the first party shall not appoint

another company or factory to sell VIP Cold Cream in the territory of Philippines without the second party's consent.

6. The second party shall enjoy the privilege of the agent forever in case they sell up to the agreed quantity every month.
7. Expense of advertisement and the repressing of false marks is for the account of the second party.
8. With the consent of the both parties concerned this agreement is functioning from the commencement of the date signed on the Agreement.
9. This agreement is made both in English literature and Chinese literature in duplicate, each of which, both English copy and Chinese copy, shall be kept by either party. In case of any discrepancy between the Chinese copy and the English copy, the Chinese copy shall govern.
10. This agreement shall be amended after both parties' consent in writing.

The first party:	The second party:
Stella Lee	Linda Moh
Representative of	Representative of
Tong Chi Industrial Corp.	Hwa Tai Trading Co., Ltd.

【譯　　文】

商業代理合約

本合約於一九八九年四月一日由 中華民國臺灣臺南金洛路四三二號唐奇工業公司（以下稱甲方）及菲列賓馬尼剌聖尼拉街三四二五號A華泰貿易公司（以下稱乙方）簽訂。

甲方願意指定乙方爲銷售甲方出品貴賓牌冷霜在菲列賓地區的獨家代理。雙方同意條件如下：

1. 甲方派乙方擔任甲方在菲列賓地區銷售貴賓牌冷霜的獨家代理。
2. 甲方供給乙方貴賓牌冷霜的成品，由乙方包裝並貼上與原樣相同的商標及標籤。
3. 乙方每月銷售不得少於五千打。
4. 如乙方六個月不能銷售雙方同意的數量，本合約可立即在任何時候作廢。
5. 在本合約有效期間，甲方未得乙方同意前，不得指派另一家公司或工廠在菲列賓地區售賣貴賓牌冷霜。
6. 倘乙方每月能銷售規定的數量，乙方有權利永遠擔任代理。
7. 廣告費及取締冒牌費用由乙方負擔。
8. 經雙方同意後，本合約自簽字日起生效。
9. 本合約用中英文簽訂，一式兩份，雙方各執一份。如中文本與英文本發生異議時，以中文本爲憑。
10. 本合約經雙方書面同意後修改之。

甲方	乙方
李優美	麥蘭德
唐奇工業公司代表	華泰貿易公司代表

第四節　加工合同

（委托電晶體收音機加工）

CONTRACT FOR PROCESSING

Messrs. Sanyan Electronic Industries, Ltd. (hereinafter called Party A)

56/28, Nishi 1-Chome, Tokyo, Japan

and

Messrs. Sanyan Electronic Industries, (TAIWAN) Ltd. (hereinafter called

Party B) 5, Yen Ping Road, Taipei, Taiwan

have mutually agreed that Party A entrusts Party B with the manufacturing of transistor radios in Taiwan with all necessary parts and materials supplied by Party A under the terms and conditions specified as follows:

1. Commodity and quantities for processing
   a. Commodity: Transistor radios
   b. Quantity: 600,000 sets in total
2. All necessary parts and materials either supplied by Party A or purchased in Taiwan by Party B are as per the list attached.
3. The processing charge for each model is as follows:
   a. 6 transistor set at US$0.25 each
   b. 8 transistor set at US$0.30 each
   c. 9 transistor set at US$0.325 each
   d. 10 transistor set at US$0.35 each
   e. 11 transistor set at US$0.375 each
   f. 12 transistor set at US$0.40 each
   g. 14 transistor set at US$0.45 each
4. a. The main parts, consumption articles and materials required for processing will be sent to Keelung by Party A and if there is any shortage or breakdown of these parts and materials, Party A should be held responsible for supplying additional replacements.

   b. Party A should pay Party B by opening L/C or T/T covering the full amount of processing charges and costs of parts, of consumption articles and of materials purchased in Taiwan by Party B one month before the shipment of the finished products concerned.

c. Party B must complete the manufacturing of all transistor radios and effect shipment within the date mutually agreed by the two parties without any delay unless any unforeseen circumstance should occur which is beyond control.

5. <u>The damage rate of parts and materials</u>

The damage rate of parts and materials in processing is 3% and such a rate of spare parts and materials should be supplied free by Party A. Should the damage rate surpass 3 %, Party B will supplement the additional required materials and parts necessary for processing.

6. Should a wrong shipment of materials and parts be sent, or should an excess of materials and parts be sent by mistake, Party B should return Party A for complete replacement(In case of shortage, Party A will make shipment of the short parts and materials for supplement) or send back the excess materials and parts at the expense of Party A.

7. All parts and materials supplied by Party A for transistor radios should be processed by Party B strictly according to the design specified by Party A without allowing any modification.

8. <u>Technical service</u>

Party A agrees to dispatch technicians to Taiwan to help training those technicians of Party B at the request of the latter at any time and allows the said technicians to remain with Party B for inspection of finished products. In such a case, Party B agrees to pay monthly salary of NT$8,000 for each person. All other expenses (including round trip tickets) will be borne by Party A.

9. All import and export procedures in connection with this contract should be taken by Party B with Chinese government concerned in

Taiwan.

10. All transistor radios processed by Party B should be shipped to the foreign buyers appointed by Party A according to instructions given by Party A in due time.

11. Other terms & conditions

    a. The trade marks of transistor radios will be supplied by Party A and should there be any illegal involvement, Party A is to be held fully responsible.

    b. All parts and materials, if necessary, purchased in Taiwan by Party B for transistor radios, their quality must measure up to standard and be appoved by Party A beforehand.

    c. For promoting export business, Party B must prepare samples of transistor radios at any time and send them to any of foreign buyers appointed by Party A. All parts and materials required for this purpose would be supplied from the spare parts sent by Party A.

12. This contract shall be made in triplicate and Party A and Party B shall, after signing all copies, retain one copy respectively and submit the other one to the Government concerned in Taiwan for registration.

SANYAN ELECTRONIC INDUSTRIES, LTD.	SANYAN ELECTRONIC INDUSTRIES (TAIWAN), LTD.
(Signature)	(Signature)
....	....
H. Tanaka, Manager	Cynthia Chang, Manager

October 18, 1989

【譯 文】

加 工 合 同

訂約人　三洋電子工業公司（以下簡稱甲方）日本東京西一町目 56/28 號

三洋電子工業(臺灣)公司（以下簡稱乙方）臺灣臺北市延平路五號

茲經雙方同意，甲方委託乙方在臺灣製造電晶體收音機，由甲方供給一切必需之零件與原料，條件如下：

1. 加工之商品及數量

　a. 商品：電晶體收音機

　b. 數量：共計六十萬件

2. 一切需用之零件與原料由甲方供給，或由乙方在臺灣購買，詳單隨附在本合同內。

3. 每種型式之加工費如下：

　a. 六個電晶體者　每件美金二角五分

　b. 八個電晶體者　每件美金三角

　c. 九個電晶體者　每件美金三角二分五厘

　d. 十個電晶體者　每件美金三角五分

　e. 十一個電晶體者　每件美金三角七分五厘

　f. 十二個電晶體者　每件美金四角

　g. 十四個電晶體者　每件美金四角五分

4. a. 加工所需之主要零件，消耗品及原料由甲方運至基隆。倘有短少或破損，甲方應負責供應補充。

　b. 甲方應於成品交運一個月前，開信用狀或電滙全部加工費以及乙方在臺灣購買之零件，消耗品及原料之費用。

　c. 除無法控制之情形外，乙方應在雙方同意之時期內完成電晶體收音機之製造及交運。

5. 零件及原料之損壞率

加工時之零件及原料損壞率定爲百分之三。此項零件及原料應由甲方免費供給。如損壞率超過百分之三，乙方應補充加工所需之零件及原料。

6. 如甲方將原料及零件誤運，或因大意而將原料及零件超運，乙方應將原件退還（倘有短少情事，甲方應予補足）或將超運部份退回，由甲方負擔費用。

7. 由甲方供給作爲製造電晶體收音機之零件及原料，乙方應嚴格按照規定之設計加工，不得變更。

8. 技術服務

甲方應乙方之請求，同意隨時派遣技術人員至臺灣，協助訓練乙方之技術人員，並允許所派之技術人員留在乙方檢驗成品。倘如此，乙方同意每人付給月薪新臺幣八千元。其他一切費用（包括來回票）概由甲方負擔。

9. 一切與本合同有關之進口及出口手續，應由乙方在臺灣向中國政府辦理。

10. 加工後之電晶體收音機，乙方應運交甲方隨時指定之外國買主。

11. 其他條件

a．電晶體之商標由甲方供給，倘有法律糾紛，甲方應負完全責任。

b．倘因必需而由乙方在臺灣購買爲製造電晶體收音機之零件及原料，其品質必須符合標準並事先經過甲方核准者。

c．爲推廣出口貿易起見，乙方應準備電晶體收音機樣品，隨時寄往甲方所指定之外國買主。所需之零件及原料，在甲方所運之零件及原料內使用。

12. 本合同一式三份，甲方與乙方在簽字後各留一份，另一份送呈在臺灣之政府備案。

三洋電子工業公司　　　　三洋電子公司（臺灣）

經理　田中太一　　　　經理　張　宜　謙

一九八九年十月十八日

第五節　建築合同

（裝設工廠冷氣機）

CONTRACT FOR WORK

This contract, made as of the day of July 1, 1989 by and between WEST ELECTRONIC CORPORATION INC. OF TAIWAN, hereinafter called as Party A and JARDINE ELECTRICAL ENGINEERING CO., LTD., hereinafter called as Party B.

WITNESSETH:

Whereas Party B will enter into a contract for work with Party A for the latter's air conditioning equipment, and the parties hereof, in consideration of the mutual covenants and agreement, do hereby agree as follows:

1. Name of work: Air conditioning equipment.
2. Location of work: Party A's newly erected plant in Nan Tse, Tainan Hsien.
3. Scope of work: As specified in the drawings and proforma invoice attached hereto.
4. Time limit for completion: The work hereof shall be commenced in compliance with the contruction work of Party A's plant, and shall be completed within thirty days after the completion of construction work of Party A's plant, including the completion of inner part of the plant when the plant be in the condition that trial run for air conditioning mechanism is possible.
5. Total price of work: One million two hundred and sixty thousand New Taiwan dollars only.

6. Payment for work:

First installment: Party A shall pay to Party B five percent of the total price for work, i.e., sixty three thousand New Taiwan dollars only, at the time when this contract be duly executed.

Second installment: Party A shall pay to Party B fifteen percent of the total price for work, i.e., one hundred and eighty nine thousand New Taiwan dollars only at the commencement of construction work of Party A's plant.

Third installment: Party A shall pay to Party B twenty percent of the total price for work, i.e., two hundred and fifty two thousand New Taiwan dollars only, upon the commencement of installation of air pipes.

Fourth installment: Party A shall pay to party B forty percent of the total price for work, i.e., five hundred and four thousand New Taiwan dollars only, at the time when the main mechanism of the air conditioning equipment each set of HMC 75F-6R-DX and 75F-8R-DX, arrive at the location of work.

Fifth installment: Party A shall pay to Party B ten percent of the total price for work, i.e., one hundred and twenty six thousand New Taiwan dollars only, upon the completion of the work hereof.

Sixth installment: Party A shall pay to Party B the balance of unpaid ten percent of the total price for work, one hundred and twenty six thousand New Taiwan dollars only, when the completed work be duly received and found qualified.

7. Penalty for default:

In the event Party B shall fail to complete the work hereof in time owing to such reasons that Party B shall be liable thereto, Party B shall pay a penalty for such default based on one out of a thousand of the total price for work per day, i.e., Party B shall pay one thousand two hundred and sixty New Taiwan dollars only for each day of such default.

8. Increase or decrease or work:

In the event Party A shall have amended, increased or decreased its project of construction work, the corresponding increase or decrease of total price shall be calculated as per the unit price specified herein by the parties hereof. In the event there shall be newly increased items of work, the parties hereof shall enter into an agreement for the unit price of such items. In the event Party B shall have to abandon any part of the completed work or any part of the materials transmitted to the location of work owing to amendment of construction project by Party A, Party A shall, after having duly received the work, pay for such completed work or such materials abandoned as per the unit price specified herein to Party B.

9. Supervision for progress:

The personnel for supervision of progress of work appointed by Party A and its representative shall have the authority to supervise the progress of work and give instructions to Party B, and Party B shall perform correctly in accordance with such instructions of Party A's personnel or representative without making any excuses.

10 Interruption of progress:

In the event Party A shall notice Party B to interrupt the progress of work beyond the causes Party B shall be liable thereto, Party B may withdraw from work and claim to Party A to pay the price by giving to Party A a list of the amount of work, materials transmitted to the location of work and the reasonable expenses and fees for checking of such list and payment of the amount of such list over the paid intallments or return of the amount of such list less than the paid installments.

11. Custody of work:

The completed work and the materials, tools, and equipment, etc. to the location of work shall be under custody of Party B after the commencement of work and before completion of work and delivery of the completed work. Unless in the event of force majeure, Party B shall be fully responsible for any and all accident or damage of the work under custody of Party B. In circumstances of force majeure, Party B may give a list of damages in detail according to the actual condition of damages and also give a projected list of price and date for recovery from such damages to Party A for check of such lists and payment therefor. In case Party A shall decide not to continue the progress of work, this contract shall be terminated under Article 10 hereof.

12. Warranty for work:

Party B shall warrant the promised condition of the work within one year after the date of receiving of the work by Party A; provided, however, Party B shall not be liable to Acts of God or misuse of the work by Party A.

13. Supplemental Provisions:

When the work is in progress, in case of any damages to Party A or any other persons caused by the Party B's fault, Party B shall be liable to compensate such damages; provided, however, Party A shall be solely liable to any such damages caused by it or other person contracted with it and when such contracted person is performing another construction work.

14. Attachments:

Attachments hereof shall be made a part of this contract and effective as any other provisions of this contract. Attachments hereof includes: (1) drawings; and (2) proforma invoice.

15. The form of this contract:

This contract shall be in duplicate to be held each by the parties hereof and shall have two copies each kept by the parties hereof for record.

16. Additional Provision:

Maintenance of mechanism: Party B shall provide free and periodic maintenance service for the air conditioning mechanism hereunder for a year after the receiving of the work by Party A by sending maintenance personnel once each month for performing maintenance.

In witness whereof, this contract has been executed by the parties as of the day and year first above written.

Party A:____________

By :____________

Address:____________

Part B :____________

By :____________

Address:____________

【譯 文】

工程合同

本合同於一九八九年七月一日由臺灣西方電子公司（以下簡稱甲方）與渣甸

電氣工程公司（以下簡稱乙方）簽訂。

乙方與甲方訂立工程合約，安裝甲方之空氣節調設備，雙方同意如下：

1. 工程名稱：空氣節調設備。
2. 工程地點：甲方在臺南縣南梓之新建工廠。
3. 工程範圍：本合約如所附之圖樣及估價單。
4. 完工期限：本工程將配合甲方之建廠工程開始，並在甲方工廠完工後三十天內完成，包括工廠內部完工可使空氣節調機械試用之時期在內。
5. 工程總價：新臺幣一百廿六萬元正。
6. 工程付款：

第一期：甲方在本合同開始實施時付給乙方工程總價百分之五，即新臺幣六萬三千元正。

第二期：甲方在建廠工程開始時付給乙方工程總價百分之十五，即新臺幣十八萬九千元正。

第三期：甲方在安裝氣管時，付給乙方工程總價百分之二十，即新臺幣二十五萬二千元正。

第四期：甲方在空氣節調設備之主要機械 HMC 75F-6R-DX 及 75F-8R-DX 每副運抵工地時，付給乙方工程總價百分之四十，即新臺幣五十萬四千元正。

第五期：甲方在工程完畢時，付給乙方工程總價百分之十，即新臺幣十二萬六千元正。

第六期：甲方在工程驗收合格後，付給乙方工程總價餘款百分之十，即新臺幣十二萬六千元正。

7. 違約罰款：倘乙方因應自行負責之原因而未準時完工，乙方應付違約罰款每天按總價千分之一計算，即新臺幣一千二百六十元正。
8. 增加或減少工程：倘甲方需修改、增加或減少其工程計劃時，總價之增減，應按雙方規定之單價計算。如係新增工程，雙方應另行協定該項新增工程之單價。倘甲方因修改原計劃而將已完工之一部份工程或已運抵工地之材料棄置，甲方在驗收後，應按雙方決定之單價付給乙方已完工之工程費用及材料費。

9. 監督進度：甲方所派定之監督工程進度人員及其代表有權監工並予乙方指導。乙方應照甲方人員或代表之指示正確施工，不得有任何藉口。

10. 工程停止：倘甲方通知乙方將工程停止，而其原因並非應由乙方負責者，乙方可停工並要求甲方依照已完成工程之數量，運抵工地之材料及其他合理費用在已交工款內增付或扣除。

11. 工程保管：在工程開始後，完工並移交前，已完工之工程以及留在工地之材料、工具、設備等等，均由乙方保管。除人力不能抵抗之災害外，乙方應對保管中之一切，負損害之全責。如遇不可抵抗之天災人禍，乙方應詳列損害實情，向甲方提出恢復原狀之價格及日期，以供甲方核對付款之用。倘甲方決定不再繼續施工，本合約應即按第十條之規定結束。

12. 工程保證：除乙方不能負責之天災或甲方使用不當之原因外，乙方應在甲方驗收後保證工程良好一年。

13. 附帶條款：在工程進行中，倘因乙方之過錯，使甲方或其他人員受到損害，乙方應負責賠償。但如損害係由於甲方或與甲方訂約之其他工程人員之過錯所造成，則應由甲方完全負責。

14. 附　　件：本合同之附件應視爲本合同之一部份與其他條款有同樣效力。附件包括(1)圖樣及(2)估價單。

15. 合同形式：本合同一式二份，甲乙雙方各執一份。另各保留二份副本供雙方存卷。

16. 外加條款：

機器之維護　乙方應在甲方驗收後，供給免費及定期對空氣節調器之保養服務一年，每月應派員來廠一次，維護正常運用。

本合同於上開日期生效。

甲　方：____________	乙　方：____________
負責人：____________	負責人：____________
地　址：____________	地　址：____________

第六節 租屋合同

（租賃三樓房屋一層）

AGREEMENT FOR TAKING A FLAT

MEMORANDUM of an undertaking, entered into this Thirtieth Day of September 1989 between C. C. Cheng of 2304 Canton Road, 2nd fl., Taipei and Janet Hsu of 999 Chi Lung Street, 3rd fl., Taipei as follows:

The said C. C. Cheng does hereby let to the said Janet Hsu a dwelling flat, situated in 999 Chi Lung Street, 3rd fl., Taipei for the term of two years certain, and so on from two years every, and so on until one month's notice to quit be given by or to either party, at the monthly rent of NT$8,650.00, payable in advance on the first day of each and every calendar month; the tenancy to commence at first day of October.

And the said C. C. Cheng does undertake to pay the land tax, the property tax, and the rates and to keep the said flat in all necessary repairs, so long as the said Janet Hsu shall continue therein. And the said Janet Hsu does undertake to take the said flat of C. C. Cheng for and at the before-mentioned term and rent, and pay all taxes except those on land or property and the rates. and to abide by the other conditions aforesaid.

Witness our hands the day and year aforesaid.

Signed: Witness: Nancy Hsu

C. C. Cheng Alexander C. Cheng

Janet Hsu

【譯　文】

租賃房屋一層合約

本合同於一九八九年九月三十日由 程志政住臺北廣州路二三〇四號二樓及徐嘉麗住臺北市吉林街九九九號三樓，共同簽訂如下：

程志政租給徐嘉麗住屋一層，位於臺北吉林街九九九號三樓，爲期二年，以後續租，每次以二年爲期，雙方如不再繼續，應於一個月前通知對方。每月租金爲新臺幣八千六百五十元，每月第一天預付。租賃自十月一日開始。

程志政應在徐嘉麗租用期內，繳付地價稅、財產稅及其他捐稅並使本樓維持必需之修繕。徐嘉麗依上述條件及租金，租賃程志政所有之房屋一層，並付除地價稅，財產稅及雜項捐稅外之一切捐稅，以及遵守上述之其他條件。

本合同於上述日期簽字。

簽字人	證人
程志政	徐　玫
徐嘉麗	程季平

第七節　抵押合同

（以股票抵借五萬元）

COLLATERAL NOTE

NT$50,000.00　　　　Taipei, April 30, 1989

On demand after date I promise to pay to the order of Mr. C. D. Kwong Fifty Thousand Dollars, with interest at 5%, for value received, without defalcation, and I have delivered herewith 1,000 shares Nan YA Co., Ltd. stock to be held as collateral security for the payment of this note and any and all notes given in renewal, substitution or extension or part renewal, substitution or extension thereof, and any other liability or indebtedness of me to the holder hereof now existing, or which may be hereafter contracted, which collaterals I hereby authorize and enpower the holder hereof, at any time, to transfer to himself and on default of payment at maturity, of either of the liabilities for which said securities are deposited as collateral security, with a view to liquidating said obligations and indebtednes, and all interest and costs thereon, to sell and transfer, in whole or in part, at public or private sale, without any previous demand or notice to me, and to apply the net proceeds, after deducting costs of sale or so much thereof as may be required, to the payment of this note and any and all such indebtedness or liability as aforesaid or either of them, at its option, in full or partially, as such proceeds may suffice, holding me still responsible for any deficiency. Further, I agree, that so often as the market price of the above securities and subsequently deposited securities shall fall to a price Insufficient to cover the indebtedness for which said collaterals have been deposited as security in amount, with ten per cent margin added thereto, I will, on demand, within twenty-four hours thereafter deposit with the holder additional security, to be approved by said holder, sufficient to cover said amount and margin; and that, in default thereof, this note and any and all indebtedness for which said securities have heen deposited as collateral shall become instantly due and payable, precisely as though said indebtedness had actually matured, and all the foregoing authority to transfer or sell and transfer said collateral shall at once be exercisable, at my risk, in case of any deficiency in realizing proceeds. All of which agreement applies with equal force to any and all securities added to the above original collaterals, as well as to any and all securities held by the holder hereof as collateral for any obligation of which this note is or may be a renewal or substitution.

In witness whereof I have hereunto set my hand and seal this Thirtieth Day of April, 1989.

Signed, sealed and delivered in the presence of

Joy S. C. Lu

Signature P. T. LEE

【譯 文】

抵押字據

新臺幣五萬元正　　　　　一九八九年四月三十日於臺北

茲承諾在訂約一年後付還鄺世第先生五萬元正，另加利息五厘，決不短欠。本人已交出南亞公司股票一千股，作爲償付這項債務的抵押品。今後本字據如有續期、更換或延長，或作局部的續期、更換或延長，此項抵押品依然具有同樣的效力。本人對於抵押品持有人，無論在目前或在將來，如倘有其他債務或拖欠，本人授權及委任此項抵押品的債權人將它移充抵押。倘或此項抵押品所負擔的任何一種債務，已經到期而未能清付，爲了清理此項債務，以及抵償各種利息與費用起見，可以將它全部或部份售出及轉讓。不論經由公開拍賣或私人售讓，事前都無須通知本人，或徵求本人的同意。此項抵押品出售後，先扣除各種應付的手續費，再將它的純收入，用來償付本單據有關的債務，及前面所述的各種債務及負欠。凡此種種債務及負欠，按售賣抵押品所得，償付其一部份或全部，完全要看此項純收入，是否足夠抵償爲準。如有不足，本人依然要繼續負責。此外，因上面抵押品的市價，時有漲落，以致抵押品的市價，和所擔保的債務，及外加百分之十保險差額不足相抵。如遇這種情形，本人同意於二十四小時之內，向此項抵押品的持有人提供經他認可的補充抵押品，以補足上述債款數目及其保險差額。如違反此項規定，本單據及以上開抵押品作擔保的債務，都作爲立即到期，立即照付。債務也就作爲實際到期。上述各項轉移或拋售及轉讓的權利，都立即生效。如變易所得的現款，不夠償債數目，本人自應負責。上面的協議，對於原有抵押品以外所附加的任何抵押品，均具有同等效力。本單據如續期或更換，提交抵押品持有人所保持的任何及所有抵押品，也同受本押據的約束。

一九八九年四月三十日　　　　立據人　李丕寅

證　人　呂素嬌

習　題

(一) 試述商業合同內應注意的事項。

(二) 試擬一簡單買賣合同。

(三) 試擬一簡單抵押合同。

第八章　應用商業文件

(一) 授 權 狀

POWER OF ATTORNEY

(1) 授 權 投 標

KNOW ALL MEN BY THESE PRESENTS THAT, we, HEMPEL's Marine Paints of 150 Loudtoftevej, DK-2800 Ayngby, Denmark, Producer of paints and related products, have, on behalf of the said company, and being fully authorized to do so, constituted and appointed, Asia Trading Co., Ltd. 64 Nanking East Road, Section 2, Taipei, the Republic of China, a true and lawful attorney for and in the name, place and stead of the said Hempel's Marine Paints, to bid on behalf of Hempel's Marine Paints, on invitation No. TF51-60032-(W) of Taiwan Fertilizer Corporation, Taiwan.

IN WITNESS WHEREOF, Hempel's Marine Paints, true and overall producers of Hempelin Silicon Aluminum 5372, have herewith set hand and seal in the city of Copenhagen, this 14th day of June 1989.

Hempel's Marine Paint

J. Nissen

【譯　文】

立授權書人製造油漆及相關產品的丹麥亨勃爾海運油漆廠，現已正式全權委派臺北市南京東路二段六十四號亞洲貿易公司爲本廠的眞正而合法的代表，參加臺灣肥料公司第 TF51-60032-(W) 號的投標。

爲證明起見，亨勃爾海運油漆廠，謹於一九八九年六月十四日簽字盖章於哥本哈根。

亨勃爾海運油漆廠

（代表）詹尼森

(2) 授權投票

ORIENTAL DEVELOPMENT CO.

Shareholder's Proxy

No. 7681
Number of shares
Common 5,000

KNOW ALL MEN BY THESE PRESENTS, that the undersigned hereby constitutes and appoints Mr. Charles Chang with power of substitution to vote at the Annual Meeting of the shareholders of the First Commercial Bank, to be held at the office of the Bank, 30 Chungking South Road, Sec. 1, Taipei on the 20th day of September, 1989 at nine o'clock a. m., according to the humber of votes that undersigned would be entitled to vote if then personally present. WITNESS the hand of the undersigned this 19th day of August, one thousand nine hundred and eighty nine.

Signed　Joseph Lee

【譯　文】

東方開發公司

股東投票授權書

第 7681 號

普通股 5,000

茲特公告，本人指派張瑞鑾先生有權於一九八九年九月二十日在臺北市重慶南路一段30號舉行之第一商業銀行股東年會中代表本人按照應有之選票投票，視同本人親投。一九八九年八月十九日

簽字 李約瑟

(二) 一般保險單投保書

April 8, 1989

Taipei

The Central Insurance Co., Ltd.

134, Wu Chang St., Taipei

Dear Sirs:

Please issue a Marine Insurance Policy and 3 copies as follows:

Policy in Name of Messrs. Far East Trading Co., Taipei ______

______

Valued at Stg £9,900 ______

Per S. S. "Wingon" ______ Sailing on 5/10/1981 ______

From Keelung ______ To London ______

Via ______ or Transhipment at Hong Kong ______

Terms All risks ______

Please state whether F. P. A. or W. A. etc.

Claims payable at Taipei ______

______

(1) stamped duplicate/s and (3) copy/ies required.

Marks	Description of Goods	Amount
AA CC Made in Taiwan	100 Cases Ladies Umbrellas	Stg. £9,900.00

Address: 85 Nanking East Rd., Taipei

Tel. 911-2583

Yours faithfully,

FAR EAST TRADING CO.

(三) 正式保險單

THE CENTRAL INSURANCE CO., LTD.

Policy No. 51318

This is to certify that this Company has insured on behalf of Far East Trading Co.

The sum of Pounds Sterling Nine Thousand Nine Hundred only.

Upon One Hundred cases ladies umbrellas

At & from Keelung to London

Ship or vessel: S. S. "Wingon"

Sailing on or about 10th May, 1989

Covering All risks

In the event of damage, to be surveyed by Oliver Survey Co. Keelung and claims payable at Taipei.

This policy is issued in Duplicate at Taipei on the 10th day of May in the year one thousand nine hundred eighty one.

The Central Insurance Co., Ltd.

....

(Manager)

(四) 商業發票

THE UNIVERSAL CO., LTD.

COMMERCIAL INVOICE

No. 171

Invoice of 1,000 pairs of rubber shoes shipped per S. S. "Oriental Pearl" sailing on October 1, 1989 from Keelung to New York.

Under your order No. 33033

Marks & No.	Package	Description	Quantity	Unit Price	Amount
/SMT/152/ New York No. 456783	Ten Cases	Rubber Shoes Color in Black	1,000 Pairs	US$2 c. i. f. New York	US$ 2,000

PACKING LIST 包裝明細單

Invoice No. 171

Shipped per S. S. "Oriental Pearl" sailing on October 1, 1989 from Keelung to New York.

Marks & No.	Package	Description	Quantity	Gross	Net
/SMT/152/ New York No. 456783	Ten Cases	Rubber Shoes Color in Black	1,000 Pairs	800 kgs.	350 kgs.

(五) 到貨通知書

THE CATHAY SHIPPING CO., LTD.

ARRIVAL NOTICE

D/O No. 1462

Date Nov. 1, 1989

Port New York

Cargo listed below is due to arrive aboard subject vessel. Delivery order can be obtained by presenting duly endorsed Original Bill of Lading to this office.

S. S. "Oriental Pearl" Voy. No. 131 Arriving on or about Nov. 5, 1989

Consignee: Min Hwa Co.

B/L No.	Marks	No. PKGS	Description of Cargo
262/5	/SMT/152/	10 Cases	Rubber Shoes Color in Black

Mail To: Min Hwa Co., 152 Fifth Avenue, New York

N. B. Terminal charges in accordance with prevailing tariff for account of cargo. Consignees are cautioned that full compliance with existing Customs Regulations must be met prior to receipt of cargo. Particulars may be obtained from any Customs Broker.

Signature

(六) 款已收到附上收條

Dear Sirs:

Your letter of March 30 is to hand, containing a cheque for $56, 500, which we have placed to your credit and for which we thank you, Enclosed is the receipt as requested for the amount.

Encl. a/s

Yours very truly,

註 ①to hand 收到 ②contain 內有

③have placed to your credit 已收入尊帳

(附件)

RECEIPT

Taipei, March 31, 1989

No. 100

RECEIVED from Mr. C. Y. Kuo

the sum of Fifty-Six Thousand Five Hundred Dollars only.

By Cheque.

THE CHINA TRADING CO., LTD.

$56, 500. 00

Per *Sam Wang*

(Revenue Stamp)

註 ①receipt 收據 ②revenue stamp 印花稅票

(七) 催收逾期帳款(第一次)

Dear Sirs:

Your account shows an overdue balance of $567. 90 which was

shipment of 100 cases of Canned Provisions, marked ◈ No. 1/100, shipped at Keelung on board M. S. "Hai Yuan" sailing for New York on Nov. 9, as per the Policy enclosed. We hand you herewith our account for $4, 200, which amount please pass to our account.

Yours very truly,

註 ①pursuant to 遵從　②instruction 指示
③as per 如同　④pass to our account 收入敝帳戶

（四）茶葉受損請付賠款

Dear Sirs:

Re: S. S. "Silver Star"

We are holders of the Policy No. 31404 issued by your Hong Kong agents on 400 cases tea valued at HK$25, 000. 00. During the voyage the ship encountered heavy weather, and in consequence 150 cases were damaged by sea water. We enclose the certificate of survey, also the Policy, which is against marine risk W. A.

Kindly adjust the claim and send us a cheque in settlement at your earliest convenience.

Yours truly,

註 ①holders 持有人　②agent 代理
③voyage 航程　④encounter 遭遇到
⑤heavy weather 惡劣天氣　⑥in consequence 結果
⑦certificate of survey 檢驗證明書　⑧W. A.　水漬險
⑨adjust 調整　⑩settlement 解決

Trucking.............................. 85.00

$15,344.84

@ 1.4906

Sea Water Damage:

1120 doz. #5813

@ $0.35/doz.

Insurable Value $392.00

Damaged at 33.33% $130.65

Non-delivery:

300 doz. #5813

@ $0.35/doz.

Insurable value.. 105.00

Total $235.65

@ 1.4906 .... $351.26

Survey fee 25.00

Amount claimed.. $376.26

內容要點

(I) 本索賠計算書包括二部分，前一部分記載與索賠有關的一般事項，後一部分記載索賠項目及計算。

(II) 索賠項目包括①海水損害(1120打) 計美金130.65元，②遺失 (300打) 計美金105元，合計美金235.65元。

(III) 由於每一美金之貨物投保 1.4906元，故 235.65 元的貨物應賠 351.26 元。此數再加公證費用 25 元，使索賠金額增至美金 376.26元。

有關保險的幾點說明

保險的種類很多，有火險、水險、人壽險、意外險、竊盜險、信用險等等，但和國際貿易發生最密切關係的却是水險 (Marine Insur-

Far Eastern Textile Co.

Taipei

Dear Sirs:

We have the honor to submit for your consideration a report on the feasibility of building a textile factory in Taoyuan as requested in your letter of March 3, 1989.

Yours truly,

James Lee & Co.

(D) "Table of contents" 也就是目錄。長的報告，超過三四頁的，應該有目錄。這是在致送書後的另一頁，內中將報告全部的內容，分主題、子題、支題等一一列出，並註明頁數。目的在於使讀者一目瞭然，易於查閱。例如:

TABLE OF CONTENTS

(E) "Summary of Report" 是將報告的內容，做一個簡單的摘要，使讀者在閱讀全文以前，先有一個概念。摘要不宜過長，以一兩頁爲最佳。敍述應力求簡潔扼要。

(F) "Text" 是報告的本文，也是報告中的最重要部份。其中分爲若干章，每章又分爲若干節。除詳細說明內容外，往往還須用圖表和統計表加以補充。文字方面，應力求通順，使人易於閱讀。切勿用艱澀冷僻的字彙，意義含混的詞句。謹守的原則是簡明、正確、清通、流暢。

(G) "Conclusion"（結論）和上面所談的"摘要"不同。結論的重心在於提出建設性的建議，供備採納。例如上面的遠東紡織公司委託一家顧問工程公司做一個報告，研究是否能在桃園設立紡織工廠。報告的結論就應該列舉可行或不可行的理由，提出建議，供遠東紡織公司參考。

(H) "Appendix" 是附錄。報告中的較大表格、設計圖、地圖、計算表，和其他參考資料都可放在附錄裏。如果是一個比較複雜的報告，也可以將附錄按性質分列為附錄（一）、附錄（二）、附錄(三)等，以清眉目。

（I）“Bibiliography”是將報告內所引用的參考書、雜誌、報紙等，分別註明刊物名稱、作者姓名、出版商名稱、出版地點、出版年份等，以利查考。比較簡單的報告，這一項是可以省免的。

（J）“Index”是索引，列在報告的最後部份，也只適用於很長的報告。目的在於便利讀者查閱名詞、定義，或其他參考資料，和一般書籍的索引性質相似。

第三節　報告的撰寫和裝訂

報告中的文字要簡潔有力，每段不宜過長，最好平均在一百五十字左右。各段應相互關連，脈絡貫通。意見方面，應力求客觀，讓讀者自已來考慮決定。

最後談到報告的印製和裝訂。 不論打字或鉛印， 不論精裝或平裝，報告一定要力求美觀悅目。報告所用的紙張必須完全白色，品質也要最佳的。標準尺寸是長十一英寸，濶八英寸半。每頁兩邊，要留至少一英寸的空位。校對時更要特別仔細，不可有拼法和標點符號的錯誤。

習　題

(一) 試論商業報告的性質。

(二) 試述商業報告的類別。

(三) 例行報告與特別報告有何不同?

(四) 何謂正式報告? 何謂非正式報告?

(五) 試舉正式報告的五個部份。

(六) 報告應如何繕正與裝訂?

附　　　　錄

(一) 商業名詞釋義

Above Par	超過票面
Acceptance	承兌
Accommodation Notes	融資票據
Account	帳，帳戶
Accrued	增加，生出利息
Accumulation	累積
Acknowledge	承認；收悉
Act	條例
Active	活動的；常常在用的
Actual Rate	實際滙價
Addressee	收信人
Administration Expense	管理費
Adjustment	調整
Advance	預付
Advice of Charge	付帳通知書
Advice of Drawing	票滙通知書
Advice of Outport Collection	代收委託書
Adressing Bank	通知銀行
Affiliate	聯號；聯行
Affix	蓋印
After date	發票後
After sight	見票後

Agency	代理行
Allotment	分配數
Allowance	津貼；折讓
Amalgamate	合併
Amendment	更改
Amortize	攤提
Analysis	分析
Annual Expenditure	歲出
Annual Revenue	歲入
Application for Authority to Purchase	委託購買證申請書
Appointed Bank	指定銀行
Appraisal	估價
Appreciation	漲價
Approved	核准
Arrears	拖欠
Article	項目，貨品
Asset	資產
Assign	轉讓
Attachment	附件
Audit	查帳
Available	有效
Bad Check	空頭支票
Bad Debt	壞帳
Balance	餘額
Balance Sheet	差額表，日計表
Bankers Association	銀行公會
Bank Draft	銀行滙票
Bankruptcy	破產
Barter	易貨
Bank's Buying Rate	銀行買價
Bank's Selling Rate	銀行賣價
Base Price	底價
Bearer	來人，持票人

Below Par	低於票面
Beneficiary	受益人，收款人
Bid bond	押標金
Bill of Exchange	滙票
Bill of Lading	提貨單
Blank endorsement	空白背書
Blue Chip	信譽卓著的股票
Board of Directors	董事會
Bona fide	出於善意；可靠的
Bond	保證；保稅稅單；債券
Books	帳簿；簿册
Bonus	花紅；福利金
Broker	掮客
Brokerage	掮佣
Budget	預算表
Bundle	梱
Business Day	營業日
Buying Offer	買盤
Cable confirmation	電報確認書
Calendar year	普通（日曆）年度
Cancel(l)ed	註銷；作廢
C. & F.	運費在內價
Capital	資金；資本
Cash	現金；付款
Cash against documents	憑單據付款
Cash on delivery	貨到付款
Cashier	出納員
Cashier's Check	銀行所開之支票
Cash With Order	先付貨款
Catalogue	目錄
Category	類目
Ceiling Price	限價
Central Bank	中央銀行

Certificate of Inspection	貨品檢驗證
Certificate of Origin	產地證明書
Certified Check	保付支票
Certified Invoice	簽證發票
Chamber of Commerce	商會
Charter	包租船（車）；執照
Chattels	動產
Chattel Mortgage	動產抵押
Check (Cheque)	支票
Checking Account	甲種戶頭（支票戶）
Check Book Stubs	支票存根簿
Check Register	支票登記簿
Chop	圖章
C. I. F.	運費保險費在內價
Cipher Telegram	密碼電報
Circular	通函
Circular Letter of Credit	巡廻信用狀
Claim	索賠
Classification	分類
Clean Bill	光票
Clean Bill of Lading	清潔提單；貨物完整之提單
Clearing	交換
Clearing-house	交換所
Close of Business	停業
Closing Rate	收盤滙率
Code Telegram	密電
Collateral Securities	抵押品
Collection	代收；託收
Commission	佣金；手續費
Commodity	商品；貨物
Compound Interest	複利
Confidential	機密
Confirmation	確認書

Confirmed Credit	確認信用證書
Confirming Bank	確認銀行
Consignee	收貨人
Consignment	寄售
Consular Invoice	領事簽證貨單
Contra Accounts	抵銷（對方）科目
Conversion Table	換算表
Copy	副本；份數
Correspondent	代理行；聯行，聯號
Counter	櫃臺
Counterfeit Notes	僞鈔
Countersign	副簽；會簽
Cover	抵償
Covering Letter	送報告或重要文件時所備之簡函
Crossed Check	劃線支票
Currency	貨幣；通貨
Current rate	當日滙率
Customs Broker	報關行
Customs Invoice	海關發票
D/A	承兌後交付單據
Data	資料
Date Draft	定期滙票
Date of Expiry	有效日期
Date of Value	起算利息日期
Days of Grace	寬限日
Debenture	公司債券
Debit Note	收款清單
Debt	債；債務
Deferred Payment	延付貨款
Delivery	交貨
Delivery Order	交貨單
Demand Draft	即期滙票
Demurrage	延期費

Denomination	票面
Deposit Book	存款簿
Depreciation	折舊
Designated Bank	指定銀行
Devaluation	貶值
Disbursement	付出
Discount	貼現；貼水；扣價
Dishonor	退票
Disposal of Proceeds	款項處置辦法
Dividend	紅利
Documentary Bill	跟單滙票
Documents	單據
D/P	付款後交付單據
Draft	滙票
Draw-back	退稅
Drawee	滙票付款人
Drawer	滙票出票人
Drawing	提款
Due date	到期日
Duplicate	第一副本
Endorsee	被背書人
Endorsement	背書
Endorsement Guaranteed	擔保背書無誤
Endorser	背書人
Entertainment	交際費
Equity	產權
Estate	財產
Estimates	估計；概算
Exchange Control	外滙管制
Ex Factory	工廠交貨價
Exhausted	用罄
Expenses	費用
Export Bill	出口滙票

Export Declaration	出口申報書
Extension	展期
Face Value	面值
F. A. S. (free alongside ship)	船邊交貨價
File	卷宗
Filing	歸檔
Fiscal Year	會計年度
Floating Rate	浮動滙率
F. O. B. (free on board)	船上交貨價
Foreign Exchange	外滙
Forwarding Agent	運輸行
Foul Bill of Lading (=Unclean Bill of Lading)	不潔提單
Franchise	免賠限度；政府特許之權利
Free of Particular Average (F. P. A.)	平安險（單獨海損不賠）
Freight Prepaid	運費先付訖
General Letter of Hypothecation	質押權利總設定書
Gross Weight	毛重
Guarantee	保證
Guarantor	保證人
Holder of Bill	持票人
Hypothecate	質押
Identification Card	身份證
In bond	關棧中交貨價
Income Tax	所得稅
Inflation	通貨膨脹
Installment	分期付款
Insurance Policy	保險單
Interest Rate	利率
Inventory	存貨清單
Irrevocable	不可撤消的
Issuing Bank	開證銀行
Item	項目

Journal	日記帳
Joint Account	用幾個人名義共同開立的戶頭
Ledger	分類帳
Legal Tender	法定貨幣
Letter of Credit	信用狀；信用證
Letter of Guarantee	保證書
Letter of Indemnity	賠償書
Liability	負債
Lien	留置權
Liquidation	清算；付淸
Loan	放款
Loose-leaf	活頁
Lump Sum	總價
Mail Transfer	信滙
Margin	保證金；定金
Marine Risk	水險
Mark	嘜頭（包裝外之標記）
Maturity	到期
Minimum Charges	最低費用
Mortgage	抵押
Mortgage Bond	抵押債券
Negotiate	讓購
Negotiating Bank	讓購滙票銀行
Nominal Account	虛帳戶
Nominal Rate	名義滙率
Notary Public	公證人
Notify	通知
Notifying Bank	通知銀行
Ocean Bill of Lading	海運提單
Ocean Freight	海運費
Offer	報售
Office Copy	辦公室留底
Official Rate	法定滙價

On Board Bill of Lading	已裝船提單
Open Account	開戶
Open Policy	未確定保險單
Opening Rate	開盤滙率
Order Bill of Lading	記名提單
Order Check	指定人支票
Ordinary Telegram	尋常電報
Original	正本
Outstanding	未付
Overdraft	透支
Over Valuation	估價過高
Packing Credit	打包貸款；包裝信用證書
Packing List	打包清單
Parity	平價
Partnership	合夥
Par Value	面值
Patent	專利權
Payee	受款人
Payer	付款人
Petty Cash	零用現金；雜費
Plain Telegram	明碼電
Pledge	質押；擔保
Position	職位
Posting	過帳
Postal Money Order	郵局滙票
Power of Attorney	委託書
Price Ceiling	高價（限最高價）
Price Floor	低價（限最低價）
Principal	本金
Priority	優先
Procuration	採購
Professional Charges	公費
Proforma invoice	估價單（並非正式報價單。通常用作申請進口證之用。）

Promissory Note	期票
Proof	證明；校樣
Pro Rata	按照比例
Protest	拒付；拒絕證書；抗議書
Quadruplicate	三副本；四份
Quality Certificate	品質證明書
Quota System	限額制
Rate of exchange	滙率
Real Estate	不動產
Received for Shipment B/L	備運提單
Red Clause	紅條款
Rediscount	重貼現
Refunds and Rebates	回扣
Register	登記簿
Reimbursement	償還；補償
Remittance	滙款
Renewal	展期
Requisition	請求單
Retire (or take up) Bill	贖票
Revenue	收入
Revocable	可取銷的
Revolving Credit	循環信用狀
R. S. D.	卸貨費用及棧租
Safe Box	保管箱
Salary	薪金
Sales Contract	售貨合約
Screen	審核
Secured Loan	抵押放款
Securities	擔保品
Shipping Documents	貨運單據
Shipping Space	船位
Shortage	短少
Shut Out	退關

Sight Draft	即期滙票
Signature Card	印鑑卡
Single Proprietorship	獨資
Slip	傳票
Sola	單張滙票（無副本的滙票）
Special Endorsement	記名背書
Statistics	統計
Stock	股票
Stock Exchange	證劵交易所
Stock Holder	股東
Stop Payment	止付
Storage	棧租
Sundry	雜項
Surveyor	鑑定人；公證人
Tariff	稅則
Telegraphic Address	電報掛號
Telegraphic Transfer	電滙
Tenor	滙票期限
Terms	條件
Through Bill of Lading	聯運提單
Tolerance	合理之伸縮
Transfer	轉讓
Transferee	受讓人
Transferor	出讓人
Transhipment	轉運；換船
Traveller's Check	旅行支票
Trust Receipt	信託收據
Turnover	週轉數；營業額
Unclean Bill of Lading	不潔提貨帳
Uncollectible Account	壞帳
Unconfirmed	未經確認的
Underwriter	保險人；下書人
Unit Price	單價

Urgent Telegram	急電
Usance Bill	遠期滙票
Usance Letter of Credit	遠期信用狀
Validity	有效期限
Voucher	傳票；憑單
Warehouse Receipt	棧單
War Risks	兵險
W. A. (With Particular Average)	水漬險（單獨海損賠償險）
Warranties	特約條款
Weight Certificate	重量證明書
Withdrawal	提款；退票
Writing-off Process	銷除法
Yard	碼（三英尺）
Yield	收益；產出

（二）商業名詞縮寫

A. B. NO.	Accepted Bill Number	進口到單編號
A/C	Account	帳戶
AC.	Acceptance	承兌
ACME	ACME Commodities Phrases Code	愛克米商品及用語密碼
Ad	Advertisement	廣告
A/D	After Date	發票後定期付款
A. F. B.	Air Freight Bill	航空提單
Agt.	Agent	代理商
AM	Amendment	修改書
Amt.	Amount	金額
A/O	Account of	進某戶帳
A/P	Authority to Purchase	委託購買證
A. P.	Account Payable	應付款
Approx.	Approximate	大約

A. R.	Account Receivable	應收款
Asst.	Assistant	助理
A/V	According to value	按値
Bal.	Balance	餘額
B/C	Bill for Collection	託收票據
B. D	Bills discounted	貼現票據
B/D	Bank Draft	銀行滙票
B/E	Bill of Exchange	滙票
B/F	Brougnt Forward	承前頁
BK	Bank	銀行
Bkg.	Banking	銀行業務
B/L	Bill of Lading	提貨單
B. N.	Bank Note	銀行鈔票
B. O.	Branch Office	分支行
B/P or B. P	Bill Purchased	買入光票
B/S or B. S.	Balance Sheet	餘額表；資產負債表
C. A.	Credit Advice	收款報單
C. A. D.	Cash aginst documents	付現交單
Canc.	Cancel	取銷
C/B	Clean Bill	光票
CBC	Central Bank of China	中央銀行
C. B. D.	Cash before delivery	付現後交貨
C. D.	Collection and Delivery	託交
C. C.	Carbon Copy	複打副本
C. C.	Chamber of Commerce	商會
Cert.	Certificte	證明書
C & F	Cost and Freight	運費在內價
C. H.	Clearing House	票據交換所
C. I.	Certificate of Insurance	保險單
C & I	Cost and Insurance	貨價及保險
C. I. F.	Cost, Insurance and Freight	運費保險費在內價
C. I. F. C.	Cost, Insurance, Freight and Commission	運費，保險費，佣金在內價

C. I. F. E.	Cost, Insurance, Freight and Exchange	運費，保險費，滙費內在價
C. I. F. I.	Cost, Insurance, Freight and Interest	運費，保險費，利息在內價
C. I. F. C. I.	Cost, Insurance, Freight, Commission and Interest	運費，保險費，佣金，利息在內價
CK	Check	支票
CL	Collection	託收
CM	Commission	佣金
C/N	Credit Note	收款通知
c/o	care of	轉交
C. O. D.	Cash on Delivery	付款交貨
C/P	Charter Party	租船契約
C. P. A.	Certified Public Accountant	會計師
Cr.	Credit	貸方
C/S	Case or Cases	箱
cts.	Cents	分
C. W. O.	Cash with Order	憑票即付
D/A	Documents against Acceptance	承兌後交單
D. A.	Debit advice	付款報單
D/D	Demand Draft	即期滙票
Dept.	Department	局，部
Disc.	Discount	貼現；折扣
DLT	Day Letter Telegram	書信電
D/N	Debit Note	付款通知
do	ditto	同上
D/O	Delivery Order	交貨單
Doz.	Dozen	打
D/P	Documents against Payment	付款交單
Dr.	Debit	借方
(60) d/s	(60) days after sight	見票後（60）日內付款
DV	Dividends	股利
EC	Error Correction	錯誤更正

Enc. or Encl.	Enclosure	附件
E. & O. E.	Errors and Omissions Excepted	有錯當查
e. g.	for instance	例如
eq.	equivalent	相等
etc.	et cetera	等等
Exp.	Export	出口
F. A. A.	Free of All Average	全損賠償
F. A. S.	Free Alongside Ship	船邊交貨
F. & D.	Freight and Demurrage	運費及延裝費
F. D.	Foreign Department or Foreign Division	國外部
F/O	in favor of	抬頭人
F. O. B.	Free on Board	船上交貨價
F. O. I.	Free of Interest	免息
F. O. R.	Free on Rail	火車上交貨價
F. O. T.	Free on Truck	卡車上交貨價
F. P. A.	Free of Particular Average	平安險
F. X.	Foreign Exchange	外滙
G/N	Guarantee of Notes	承諾保證
H. O.	Head Office	總行
I/C	Inward Collection	進口託收
i. e.	that is	就是
Imp.	Import	進口
IN	Interest	利息
IOU	I owe you	欠條
Insp.	Inspection	檢驗
inst.	Instant	本月份
Insur.	Insurance	保險
Inv.	Invoice	發票
I/P	Insurance Policy	保險單
I/R	Inward Remittance	滙入滙款
J/A	Joint Account	共同帳戶
Kg.	Kilogram	公斤

L/A	Letter of Authorization	授權書
lbs.	Pounds	磅
L/C	Letter of Credit	信用狀
L/H	General Letter of Hypothecation	質押權利總設定書
L/I	Letter of Indemnity	賠償保證書
L. T.	Long Ton	長噸（2,240磅）
L/U	Letter of Undertaking	承諾書
Memo	Memorandum	備忘錄
M. I.	Marine Insurance	海險
M/N	Minimum	最低額
M/T	Mail Transfer	信滙
M. T.	Metric Ton	公噸
N. B.	Nota bene (=note well)	注意
O/C	Outward Collection	出口託收
OD.	Overdraft	透支
O. P.	Open Policy	預定保單
O/R	Outward Remittance	滙出滙款
OZ.	Ounce	英兩；盎斯
P/A	Procurement Authorization	採購授權書
P. C.	Percent	百分數（%）
Pkg.	Package	包裹
Pd.	Paid	已付
Per Pro	Per Procuration (by power of authority)	代理
P. F.	Pro Forma	預估
P. & I.	Protection and Indemnity	意外險
P/N	Promissory Note	期票；本票
P. O. D.	Pay on Delivery	發貨付款
P/O	Payment Order	支付命令
P. O. B.	Post Office Box	郵政信箱
P. S.	Post Script	附言
P. T. O.	Please turn over	請反閱後面
Ref.	Reference	摘由；參考

RM	Remittance	滙款
R. O.	Remittance Order	滙款委託書
SD.	Sundries	雜項
S/D	Sight Draft	見票後即付滙票
SE.	Securities	抵押品
S. S.	Steamship	輪船
T. P. N. D.	Theft, Pilferage and Non-delivery	盜竊及不能交貨險
T/R	Trust Receipt	信託收據
T/T	Telegraphic Transfer	電滙
W. A.	With Particular Average	水漬險（單獨海損照賠）

(三)美國各州州名簡寫

Abbreviations of names of states in the United States

Ala.	Alabama	N. Dak.	North Dakota
Ariz.	Arizona	N. H.	New Hampshire
Ark.	Arkansas	N. J.	New Jersey
Calif.	California	N. Mex.	New Mexico
Colo.	Colorado	N. Y.	New York
Conn.	Connecticut	Nebr.	Nebraska
Del.	Delaware	Nev.	Nevada
Fla.	Flordia	Okla.	Oklahoma
Ga.	Georgia	Oreg.	Oregon
Ill.	Illinois	Pa.	Pennsylvania
Ind.	Indiana	R. I.	Rhode Island
Kans.	Kansas	S. C.	South Carolina
Ky.	Kentucky	S. Dak.	South Dakota
La.	Louisiana	Tenn.	Tennessee
Mass.	Massachusetts	Tex.	Texas
Md.	Maryland	Va.	Virginia
Mich.	Michigan	Vt.	Vermont
Minn.	Minnesota	Wash.	Washington
Miss.	Mississippi	Wis.	Wisconsin
Mo.	Missouri	W. Va.	West Virginia
Mont.	Montana	Wyo.	Wyoming
N. C.	North Carolina		

The names of the following states are not ordinarily abbreviated:

Alaska Hawaii Idaho Iowa
Maine Ohio Utah

The name of the District of Columbia is almost never written out in full; in fact, it is usually referred to as "D. C.,"using the abbreviation as the name. The postal address of the capital city is "Washington, D.C."

三民大專用書書目

國父遺教

書名	作者		學校
國父思想	涂子麟	著	中山大學
國父思想	周世輔	著	前政治大學
國父思想新論	周世輔	著	前政治大學
國父思想要義	周世輔	著	前政治大學

法　律

書名	作者		學校
中國憲法新論	薩孟武	著	前臺灣大學
中國憲法論	傅肅良	著	中興大學
中華民國憲法論	管歐	著	東吳大學
中華民國憲法逐條釋義(一)～(四)	林紀東	著	臺灣大學
比較憲法	鄒文海	著	前政治大學
比較憲法	曾繁康	著	臺灣大學
美國憲法與憲政	荆知仁	著	政治大學
國家賠償法	劉春堂	著	輔仁大學
民法概要	鄭玉波	著	臺灣大學
民法概要	董世芳	著	實踐學院
民法總則	鄭玉波	著	臺灣大學
判解民法總則	劉春堂	著	輔仁大學
民法債編總論	鄭玉波	著	臺灣大學
判解民法債篇通則	劉春堂	著	輔仁大學
民法物權	鄭玉波	著	臺灣大學
判解民法物權	劉春堂	著	輔仁大學
民法親屬新論	黃宗樂	等著	臺灣大學
民法繼承新論	黃宗樂	等著	臺灣大學
商事法論	張國鍵	著	臺灣大學
商事法要論	梁宇賢	著	中興大學
公司法	鄭玉波	著	臺灣大學
公司法論	柯芳枝	著	臺灣大學

書名	作者		單位
公司法論	梁宇賢	著	中興大學
票據法	鄭玉波	著	臺灣大學
海商法	鄭玉波	著	臺灣大學
海商法論	梁宇賢	著	中興大學
保險法論	鄭玉波	著	臺灣大學
民事訴訟法釋義	石志泉 原著 楊建華 修訂		輔仁大學
破產法	陳榮宗	著	臺灣大學
破產法論	陳計男	著	行政法院
刑法總整理	曾榮振	著	臺中地院
刑法總論	蔡墩銘	著	臺灣大學
刑法各論	蔡墩銘	著	臺灣大學
刑法特論（上）（下）	林山田	著	政治大學
刑事政策（修訂版）	張甘妹	著	臺灣大學
刑事訴訟法論	黃東熊	著	中興大學
刑事訴訟法論	胡開誠	著	臺灣大學
行政法（改訂版）	林紀東	著	臺灣大學
行政法	張家洋	著	政治大學
行政法之基礎理論	城仲模	著	中興大學
犯罪學	林山田	等著	政治大學等
監獄學	林紀東	著	臺灣大學
土地法釋論	焦祖涵	著	東吳大學
土地登記之理論與實務	焦祖涵	著	東吳大學
引渡之理論與實踐	陳榮傑	著	外交部
國際私法	劉甲一	著	臺灣大學
國際私法新論	梅仲協	著	前臺灣大學
國際私法論叢	劉鐵錚	著	政治大學
現代國際法	丘宏達	等著	馬利蘭大學等
現代國際法基本文件	丘宏達	編	馬利蘭大學
平時國際法	蘇義雄	著	中興大學
中國法制史	戴炎輝	著	臺灣大學
法學緒論	鄭玉波	著	臺灣大學
法學緒論	孫致中	著	各大專院校

強制執行法　陳榮宗 著　臺灣大學
法院組織法論　管歐 著　東吳大學

政治・外交

政治學　薩孟武 著　前臺灣大學
政治學　鄒文海 著　前政治大學
政治學　曹伯森 著　陸軍官校
政治學　呂亞力 著　臺灣大學
政治學概要　張金鑑 著　政治大學
政治學方法論　呂亞力 著　臺灣大學
政治理論與研究方法　易君博 著　政治大學
公共政策概論　朱志宏 著　臺灣大學
公共政策　曹俊漢 著　臺灣大學
公共政策　朱志宏 著　臺灣大學
公共關係　王德馨 等著　交通大學
中國社會政治史(一)～(四)　薩孟武 著　前臺灣大學
中國政治思想史　薩孟武 著　前臺灣大學
中國政治思想史（上）（中）（下）　張金鑑 著　政治大學
西洋政治思想史　張金鑑 著　政治大學
西洋政治思想史　薩孟武 著　前臺灣大學
中國政治制度史　張金鑑 著　政治大學
比較主義　張亞澐 著　政治大學
比較監察制度　陶百川 著　國策顧問
歐洲各國政府　張金鑑 著　政治大學
美國政府　張金鑑 著　政治大學
地方自治概要　管歐 著　東吳大學
國際關係——理論與實踐　朱張碧珠 著　臺灣大學
中美早期外交史　李定一 著　政治大學
現代西洋外交史　楊逢泰 著　政治大學

行政・管理

行政學（增訂版）　張潤書 著　政治大學
行政學　左潞生 著　中興大學
行政學新論　張金鑑 著　政治大學

行政管理學	傅肅良　著	中興大學
行政生態學	彭文賢　著	中興大學
各國人事制度	傅肅良　著	中興大學
考銓制度	傅肅良　著	中興大學
交通行政	劉承漢　著	成功大學
組織行為管理	龔平邦　著	前逢甲大學
行為科學概論	龔平邦　著	前逢甲大學
行為科學與管理	徐木蘭　著	臺灣大學
組織行為學	高尚仁　等著	香港大學
組織原理	彭文賢　著	中興大學
實用企業管理學	解宏賓　著	中興大學
企業管理	蔣靜一　著	逢甲大學
企業管理	陳定國　著	臺灣大學
國際企業論	李蘭甫　著	中文大學
企業政策	陳光華　著	交通大學
企業概論	陳定國　著	臺灣大學
管理新論	謝長宏　著	交通大學
管理概論	郭崑謨　著	中興大學
管理個案分析	郭崑謨　著	中興大學
企業組織與管理	郭崑謨　著	中興大學
企業組織與管理（工商管理）	盧宗漢　著	中興大學
現代企業管理	龔平邦　著	前逢甲大學
現代管理學	龔平邦　著	前逢甲大學
事務管理手册	新聞局　著	
生產管理	劉漢容　著	成功大學
管理心理學	湯淑貞　著	成功大學
管理數學	謝志雄　著	東吳大學
品質管理	戴久永　著	交通大學
可靠度導論	戴久永　著	交通大學
人事管理（修訂版）	傅肅良　著	中興大學
作業研究	林照雄　著	輔仁大學
作業研究	楊超然　著	臺灣大學
作業研究	劉一忠　著	舊金山州立大學

系統分析	陳進 著	前聖瑪麗大學

社會

社會學	蔡文輝 著	印第安那大學
社會學	龍冠海 著	前臺灣大學
社會學	張華葆 主編	東海大學
社會學理論	蔡文輝 著	印第安那大學
社會學理論	陳秉璋 著	政治大學
社會心理學	劉安彥 著	傑克遜州立大學
社會心理學	張華葆 著	東海大學
社會心理學	趙淑賢 著	安柏拉校區
社會心理學理論	張華葆 著	東海大學
政治社會學	陳秉璋 著	政治大學
醫療社會學	廖榮利 等著	臺灣大學
組織社會學	張苙雲 著	臺灣大學
人口遷移	廖正宏 著	臺灣大學
社區原理	蔡宏進 著	臺灣大學
人口教育	孫得雄 編著	東海大學
社會階層化與社會流動	許嘉猷 著	臺灣大學
社會階層	張華葆 著	東海大學
西洋社會思想史	張承漢 等著	臺灣大學
中國社會思想史（上）（下）	張承漢 著	臺灣大學
社會變遷	蔡文輝 著	印第安那大學
社會政策與社會行政	陳國鈞 著	中興大學
社會福利行政（修訂版）	白秀雄 著	臺灣大學
社會工作	白秀雄 著	臺灣大學
社會工作管理	廖榮利 著	臺灣大學
團體工作：理論與技術	林萬億 著	臺灣大學
都市社會學理論與應用	龍冠海 著	前臺灣大學
社會科學概論	薩孟武 著	前臺灣大學
文化人類學	陳國鈞 著	中興大學

書名	作者		單位
勞工問題	陳國鈞	著	中興大學
少年犯罪心理學	張華葆	著	東海大學
少年犯罪預防及矯治	張華葆	著	東海大學

教　育

書名	作者		單位
教育哲學	賈馥茗	著	臺灣師大
教育哲學	葉學志	著	彰化教院
普通教學法	方炳林	著	前臺灣師大
各國教育制度	雷國鼎	著	臺灣師大
教育心理學	溫世頌	著	傑克遜州立大學
教育心理學	胡秉正	著	政治大學
教育社會學	陳奎憙	著	臺灣師大
教育行政學	林文達	著	政治大學
教育行政原理	黃文輝	主譯	臺灣師大
教育經濟學	蓋浙生	著	臺灣師大
教育經濟學	林文達	著	政治大學
工業教育學	袁立錕	著	彰化教院
技術職業教育行政與視導	張天津	著	臺灣師大
技職教育測量與評鑑	李大偉	著	臺灣師大
高科技與技職教育	楊啟棟	著	臺灣師大
工業職業技術教育	陳昭雄	著	臺灣師大
技術職業教育教學法	陳昭雄	著	臺灣師大
技術職業教育辭典	楊朝祥	編著	臺灣師大
技術職業教育理論與實務	楊朝祥	著	臺灣師大
工業安全衛生	羅文基	著	
人力發展理論與實施	彭台臨	著	臺灣師大
職業教育師資培育	周談輝	著	臺灣師大
家庭教育	張振宇	著	淡江大學
教育與人生	李建興	著	臺灣師大
當代教育思潮	徐南號	著	臺灣師大
比較國民教育	雷國鼎	著	臺灣師大
中等教育	司琦	著	政治大學
中國教育史	胡美琦	著	文化大學

中國現代教育史　鄭世興 著　臺灣師大
中國大學教育發展史　伍振鷟 著　臺灣師大
中國職業教育發展史　周談輝 著　臺灣師大
社會教育新論　李建興 著　臺灣師大
中國社會教育發展史　李建興 著　臺灣師大
中國國民教育發展史　司琦 著　政治大學
中國體育發展史　吳文忠 著　臺灣師大
如何寫學術論文　宋楚瑜 著　臺灣大學
論文寫作研究　段家鋒 等著　政戰學校等

心理學

心理學　劉安彥 著　傑克遜州立大學
心理學　張春興 等著　臺灣師大等
人事心理學　黃天中 著　淡江大學
人事心理學　傅肅良 著　中興大學

經濟・財政

西洋經濟思想史　林鐘雄 著　臺灣大學
歐洲經濟發展史　林鐘雄 著　臺灣大學
比較經濟制度　孫殿柏 著　政治大學
經濟學原理（增訂新版）　歐陽勛 著　政治大學
經濟學導論　徐育珠 著　南康涅狄克州立大學
經濟學概要　歐陽勛 等著　政治大學
通俗經濟講話　邢慕寰 著　前香港大學
經濟學（增訂版）　陸民仁 著　政治大學
經濟學概論　陸民仁 著　政治大學
國際經濟學　白俊男 著　東吳大學
國際經濟學　黃智輝 著　東吳大學
個體經濟學　劉盛男 著　臺北商專
總體經濟分析　趙鳳培 著　政治大學
總體經濟學　鐘甦生 著　西雅圖銀行
總體經濟學　張慶輝 著　政治大學
總體經濟理論　孫震 著　臺灣大學

數理經濟分析	林大侯 著	臺灣大學
計量經濟學導論	林華德 著	臺灣大學
計量經濟學	陳正澄 著	臺灣大學
經濟政策	湯俊湘 著	中興大學
合作經濟概論	尹樹生 著	中興大學
農業經濟學	尹樹生 著	中興大學
工程經濟	陳寬仁 著	中正理工學院
銀行法	金桐林 著	華南銀行
銀行法釋義	楊承厚 著	銘傳管理學院
商業銀行實務	解宏賓 編著	中興大學
貨幣銀行學	何偉成 著	中正理工學院
貨幣銀行學	白俊男 著	東吳大學
貨幣銀行學	楊樹森 著	文化大學
貨幣銀行學	李穎吾 著	臺灣大學
貨幣銀行學	趙鳳培 著	政治大學
現代貨幣銀行學	柳復起 著	新南威爾斯大學
現代國際金融	柳復起 著	新南威爾斯大學
國際金融理論與制度（修訂版）	歐陽勛等編著	政治大學
金融交換實務	李麗 著	中央銀行
財政學	李厚高 著	逢甲大學
財政學（修訂版）	林華德 著	臺灣大學
財政學原理	魏萼 著	臺灣大學
商用英文	張錦源 著	政治大學
商用英文	程振粵 著	臺灣大學
貿易契約理論與實務	張錦源 著	政治大學
貿易英文實務	張錦源 著	政治大學
信用狀理論與實務	蕭啟賢 著	輔仁大學
信用狀理論與實務	張錦源 著	政治大學
國際貿易	李穎吾 著	臺灣大學
國際貿易實務詳論	張錦源 著	政治大學
國際貿易實務	羅慶龍 著	逢甲大學

國際貿易理論與政策（修訂版）	歐陽勛等編著	政治大學
國際貿易政策概論	余德培 著	東吳大學
國際貿易論	李厚高 著	逢甲大學
國際商品買賣契約法	鄧越今 編著	外貿協會
國際貿易法概要	于政長 著	東吳大學
國際貿易法	張錦源 著	政治大學
外匯投資理財與風險	李麗 著	中央銀行
外匯、貿易辭典	于政長 編著 張錦源 校訂	東吳大學 政治大學
貿易實務辭典	張錦源 編著	政治大學
貿易貨物保險（修訂版）	周詠棠 著	中央信託局
貿易慣例	張錦源 著	政治大學
國際匯兌	林邦充 著	政治大學
國際行銷管理	許士軍 著	新加坡大學
國際行銷	郭崑謨 著	中興大學
行銷管理	郭崑謨 著	中興大學
海關實務（修訂版）	張俊雄 著	淡江大學
美國之外匯市場	于政長 譯	東吳大學
保險學（增訂版）	湯俊湘 著	中興大學
人壽保險學（增訂版）	宋明哲 著	德明商專
人壽保險的理論與實務	陳雲中 編著	臺灣大學
火災保險及海上保險	吳榮清 著	文化大學
市場學	王德馨 等著	中興大學
行銷學	江顯新 著	中興大學
投資學	龔平邦 著	前逢甲大學
投資學	白俊男 等著	東吳大學
海外投資的知識	葉雲鎮 等譯	
國際投資之技術移轉	鍾瑞江 著	東吳大學

會計・統計・審計

銀行會計（上）（下）	李兆萱 等著	臺灣大學等
初級會計學（上）（下）	洪國賜 著	淡水工商
中級會計學（上）（下）	洪國賜 著	淡水工商
中等會計（上）（下）	薛光圻 等著	西東大學等

會計辭典	龍毓聃 譯	
會計學（上）（下）	幸世間 著	臺灣大學
會計學題解	幸世間 著	臺灣大學
成本會計（上）（下）	洪國賜 著	淡水工商
成本會計	盛禮約 著	淡水工商
政府會計	李增榮 著	政治大學
政府會計	張鴻春 著	臺灣大學
稅務會計	卓敏枝 等著	臺灣大學等
財務報表分析	洪國賜 等著	淡水工商等
財務報表分析	李祖培 著	中興大學
財務管理	張春雄 著	政治大學
財務管理（增訂新版）	黃柱權 著	政治大學
商用統計學（修訂版）	顏月珠 著	臺灣大學
商用統計學	劉一忠 著	舊金山州立大學
統計學（修訂版）	柴松林 著	政治大學
統計學	劉南溟 著	前臺灣大學
統計學	張浩鈞 著	臺灣大學
統計學	楊維哲 著	臺灣大學
統計學	顏月珠 著	臺灣大學
統計學題解	顏月珠 著	臺灣大學
推理統計學	張碧波 著	銘傳管理學院
應用數理統計學	顏月珠 著	臺灣大學
統計製圖學	宋汝濬 著	臺中商專
統計概念與方法	戴久永 著	交通大學
審計學	殷文俊 等著	政治大學
商用數學	薛昭雄 著	政治大學
商用數學（含商用微積分）	楊維哲 著	臺灣大學
線性代數（修訂版）	謝志雄 著	東吳大學
商用微積分	何典恭 著	淡水工商
微積分	楊維哲 著	臺灣大學
微積分（上）（下）	楊維哲 著	臺灣大學
大二微積分	楊維哲 著	臺灣大學